Rhoddy A. Viveros Muñoz

Speech perception in complex acoustic environments

Evaluating moving maskers using virtual acoustics

Logos Verlag Berlin GmbH

Aachener Beiträge zur Akustik

Editors:
Prof. Dr. rer. nat. Michael Vorländer
Prof. Dr.-Ing. Janina Fels
Institute of Technical Acoustics
RWTH Aachen University
52056 Aachen
www.akustik.rwth-aachen.de

Bibliographic information published by the Deutsche Nationalbibliothek

The Deutsche Nationalbibliothek lists this publication in the Deutsche Nationalbibliografie; detailed bibliographic data are available in the Internet at http://dnb.d-nb.de .

D 82 (Diss. RWTH Aachen University, 2019)

ISBN 978-3-8325-4963-3
ISSN 2512-6008
Vol. 31

Logos Verlag Berlin GmbH
Comeniushof, Gubener Str. 47,
D-10243 Berlin
Tel.:   +49 (0)30 / 42 85 10 90
Fax: +49 (0)30 / 42 85 10 92
http://www.logos-verlag.de

Speech perception in complex acoustic environments

Evaluating moving maskers using virtual acoustics

Von der Fakultät für Elektrotechnik und Informationstechnik der
Rheinischen-Westfälischen Technischen Hochschule Aachen
zur Erlangung des akademischen Grades eines

DOKTORS DER INGENIEURWISSENSCHAFTEN

genehmigte Dissertation

vorgelegt von

M.Sc.

Rhoddy Angel Viveros Muñoz

aus Linares, Chile

Berichter:

Universitätsprofessorin Dr.-Ing. Janina Fels
Universitätsprofessor Dr. Steven van de Par

Tag der mündlichen Prüfung: 24.05.2019

Diese Dissertation ist auf den Internetseiten der Hochschulbibliothek online verfügbar.

Abstract

Listeners with hearing impairments have difficulties understanding speech in the presence of background noise. Although prosthetic devices like hearing aids may improve the hearing ability, listeners with hearing impairments still complain about their speech perception in the presence of noise. Pure-tone audiometry gives reliable and stable results, but the degree of difficulties in spoken communication cannot be determined. Therefore, speech-in-noise tests measure the hearing impairment in complex scenes and are an integral part of the audiological assessment.

In everyday acoustic environments, listeners often need to resolve speech targets in mixed streams of distracting noise sources. This specific acoustic environment was first described as the "cocktail party" effect and most research has concentrated on the listener's ability to understand speech in the presence of another voice or noise, as a masker. Speech reception threshold (SRT) for different spatial positions of the masker(s) as a measure of speech intelligibility has been measured. At the same time, the benefit of the spatial separation between speech target and masker(s), known as spatial release from masking (SRM), was largely investigated. Nevertheless, previous research has been mainly focused on studying only stationary sound sources. However, in real-life listening situations, we are confronted with multiple moving sound sources such as a moving talker or a passing vehicle. In addition, head movements can also lead to moving sources. Thus, the present thesis deals with quantifying the speech perception in noise of moving maskers under different complex acoustic scenarios using virtual acoustics.

In the first part of the thesis, the speech perception with a masker moving both away from the target position and toward the target position was analyzed. From these measures, it was possible to assess the spatial separation benefit of a moving masker. Due to the relevance of spatial separation on intelligibility, several models have been created to predict SRM for stationary maskers. Therefore, this thesis presents a comparative analysis between moving maskers and previous models for stationary maskers to investigate if the models are able to predict SRM of maskers in movement. Due to the results found in this thesis, a new mathematical model to predict SRM for moving maskers is presented.

In real-world scenarios, listeners often move their head to identify the sound source of interest. Thus, this thesis also investigates if listeners use their head movements to maximize the intelligibility in an acoustic scene with a masker in movement. A higher SRT (worse intelligibility) was found with the head movement condition than in the condition without head movement. Also, the use of an individual head-related transfer function (HRTF) was evaluated in comparison to an artificial-head HRTF. Results showed significant differences between individual and artificial HRTF, reflecting higher SRTs (worse intelligibility) for artificial HRTF than individual HRTF.

The room acoustics is another relevant factor that affects speech perception in noise. For maskers in movement, an analysis comparing different masker trajectories (circular and radial movements) among different reverberant conditions (anechoic, treated and untreated room) is presented. This analysis was carried out within two groups of subjects: young and elderly normal hearing. For circular and radial movements, the elderly group showed greater difficulties in understanding speech with moving masker than stationary masker.

To summarize, several cases show significant differences between the speech perception of maskers in movement and stationary maskers. Therefore, a listening test that presents moving maskers could be relevant in the clinical assessment of speech perception in noise closer to real situations.

Contents

1 Introduction   **1**

2 Fundamentals of Auditory Perception   **4**

2.1 Normal Hearing and Auditory Sensation 4

2.2 Spatial Hearing 9

    2.2.1 Head-related transfer function 12

    2.2.2 Binaural cues 13

    2.2.3 Monaural cues 16

    2.2.4 Binaural sluggishness 16

    2.2.5 Spatial unmasking 17

    2.2.6 Head movements 19

2.3 Binaural Reproduction Technique 20

    2.3.1 Binaural recording 21

    2.3.2 Binaural synthesis 22

    2.3.3 Headphone transfer function 22

    2.3.4 Individual head-related transfer function 23

3 Review of Speech-in-Noise Perception   **25**

3.1 Speech Reception Threshold 26

3.2 Reliable Speech-in-Noise Tests 27

3.3 Spatial Release From Masking 32

3.4 Predictive Auditory Processing Models 34

3.5 How Reverberation Affects Speech-in-Noise Perception 35

3.6 Studies on Moving Sound Sources 38

4 Experimental Setup   **40**

4.1 Acoustic Virtual Reproduction Software 40

4.2 Dynamic Binaural Reproduction 42

5 Digit Triplet Test   **45**

5.1 Construction 45

    5.1.1 Word selection 46

5.1.2 Speaker 46
5.1.3 Recording 47
5.1.4 Resynthesis 47
5.1.5 Masking noise 47
5.2 Optimization 48
5.3 Evaluation 49
5.4 Sequence presentation 50

6 Dynamic Speech-in-Noise Test   52
6.1 Experimental Methodology 54
6.2 Results 55
6.3 Discussion and Conclusions 57

7 Dynamic SRM: Binaural and Monaural Contributions   59
7.1 Basic Concepts 61
7.2 Experimental Methodology 63
7.3 Results 66
7.4 Discussion and Conclusions 71

8 Listeners Head Movements in a Dynamic Speech-in-Noise Test   72
8.1 Experimental Methodology 73
8.2 Results 79
      8.2.1 Speech reception threshold 79
      8.2.2 Spatial release from masking 81
      8.2.3 Stationary vs. moving masker 83
8.3 Discussion 86
      8.3.1 Head movement behavior 86
      8.3.2 Stationary vs. moving masker comparison 87
      8.3.3 Benefit of head movements on intelligibility 87
8.4 The Role of Individual HRTF 91
      8.4.1 Procedure 93
      8.4.2 Results 95
8.5 Discussion and Conclusions 98

9 Assessment of Different Reverberant Conditions in Young and Elderly
Subjects at Circular and Radial Masker Conditions   100
9.1 Experimental Methodology 103
      9.1.1 Virtual stimuli 103
      9.1.2 Apparatus and procedure 110
9.2 Results 111
      9.2.1 Circular conditions 112

9.2.2   Radial conditions 115

9.2.3   IACC results 120

9.3   Discussion and Conclusions 120

9.3.1   Effect of moving masker in circular conditions 122

9.3.2   Effect of moving masker in radial conditions 122

9.3.3   Effect of age in circular conditions 123

9.3.4   Effect of age in radial conditions 125

9.3.5   Conclusion 126

10 Conclusion and Outlook   **128**

List of Figures   **135**

List of Tables   **143**

Curriculum Vitae   **166**

III

1

Introduction

Listeners with hearing impairments have difficulties understanding speech in the presence of background noise. Although prosthetic devices such as hearing aids and cochlear implants may improve the hearing ability, listeners with hearing impairments still complain about their speech perception in the presence of noise [110]. Even when basic tonal audiometry is simple, easy to perform and gives reliable and stable results, it only gives a cursory idea of the degree of difficulty in spoken communication caused by hearing loss because it does not assess the ability to understand speech [110]. Therefore, the use of speech-in-noise tests to measure hearing loss in complex scenes is an integral part of a patient's audiological study [202]. Testing speech-in-noise capacities is also important in evaluating and optimizing the fitting parameters of hearing aids and cochlear implants.

Speech perception refers to the ability to understand speech to communicate effectively in everyday situations. As simple as it may seem, in some situations this could be very difficult due to the presence of masking sounds always surrounding us. This is of the utmost importance since our communication depends on our capacity to understand each other and the lack of a good communication (e.g. due to hearing loss) could lead us to psychological problems among others [176, 195, 216].

In everyday acoustic environments, we are confronted with multiple sound sources that disturb our speech perception. This specific acoustic environment was first described as the "cocktail-party" phenomenon by Cherry [53]. Most research on the cocktail-party problem has concentrated on listener's ability to understand speech in the presence of another voice or a noise, as a masker [67, 104, 149, 204]. The researchers measured speech reception threshold (SRT) for different spatial positions of the masker(s) as a measure of speech perception. At the same time, the benefit of the spatial separation between speech target and masker(s), known as spatial release from masking (SRM), was largely investigated [20, 34, 39, 40, 65, 121].

Nevertheless, previous research has been largely focused on studying only stationary sound sources, but as we know in real-life listening situations we are confronted with multiple stationary and moving sound sources that disturb our speech perception. In natural acoustic scenes, conversations may become very difficult to understand in the presence of moving maskers sources.

Since masker noises in real-world listening are not always stationary, such as a moving talker or a passing vehicle, this thesis deals with quantifying SRT and SRM of moving maskers through virtual sound sources presented binaurally via headphones. For the binaural reproduction, a set of head-related transfer functions (HRTFs), measured from the ITA artificial head [28, 214], was convolved with a speech stimulus to be rendered in free-field and reverberant conditions. All free-field virtual sound sources were simulated using the real-time software Virtual Acoustics (VA) [114]. For the reverberant simulations, the software library RAVEN [217, 218] was used. Both softwares were developed at the Institute of Technical Acoustics (ITA), RWTH Aachen University.

In chapter 6 a virtual acoustic environment was simulated to assess moving maskers, attempting to address the question: what is the amount of SRM of a moving masker? An SRM analysis with maskers moving on different trajectories could therefore bring insight into dynamic binaural speech intelligibility.

Due to the relevance of SRM analysis, several models have been created to predict SRM for diverse spatial positions of the masker and for different masker types. However, so far, none of the models takes into account maskers in movement. In chapter 7 a comparative analysis to know if previous models for stationary maskers are able to predict SRM of maskers in movement is presented. Because of the results found in this thesis, a new predictive model for moving maskers is presented.

For clarification, the terms "dynamic" and "static" are used to describe two modes of binaural reproduction, with and without listener's head movement in the virtual acoustic scene, respectively (real-time reproduction); whereas "moving" and "stationary" are reserved for describing the masker trajectories.

In real-world environments, listeners often orient their head to look for the sound source of interest. While head movement has been shown to improve sound localization accuracy [207, 230, 234], how it affects performance in SRM remains largely unknown. In chapter 8 a study is presented comparing static and dynamic reproduction to investigate if listeners could use head movements to try to maximize their speech-in-noise perception in an acoustic scene with a masker in movement. At the same time, it is known that many factors affect SRM, such as measurement paradigm, head movements, room acoustics, masker type and its spatial distribution [34]. For a virtual reproduction, another factor could also affect the SRM: individual HRTF. Therefore, subjects with individual HRTFs were compared to the use of artificial-head HRTF to clarify how individual HRTFs affect the speech perception in virtual environments.

Another relevant factor that affects SRM is room acoustics, and for maskers in movements, little has been studied. For that reason, in chapter 9, an analysis to compare different masker conditions (circular and radial movements) among different reverberant conditions (anechoic, treated and untreated) is presented. This analysis was carried out within two groups of subjects: young and elderly (no hearing aid users) subjects.

2

Fundamentals of Auditory Perception

In a healthy auditory system, the auditory perception can be defined as the ability to obtain and interpret acoustic information about our surroundings, using the pressure fluctuations in the air (i.e., sound waves) that reach the ears through audible frequencies (between 20 and 20.000 Hz). The auditory perception assists us in many social situations deciphering what people are saying, recognizing voices, and emotional states in just a few moments. However, this seemingly simple task is actually very complex and requires the use of several brain areas that are specialized in auditory perception [173]. In the hearing process, the information is carried by pressure variations in the air (sound waves) and then is converted in a way that can be used by the brain in the form of electrical activity in nerve cells or neurons.

The system responsible for the perception of sound waves is the auditory system which requires a series of processes in order to perceive the sounds around us. When an object produces a sound (auditory stimulus), the waves produced by this action are transmitted by the air (or other means) with enough intensity to reach our ears. It is also necessary for the sound to be within the audible frequency range. If these two requirements are fulfilled, the brain is able to detect where the object is and even tell if it is moving.

2.1 Normal Hearing and Auditory Sensation

Defining the range of hearing considered "normal hearing" is difficult due to the high variability of the hearing threshold.

The first attempts in quantifying hearing sensitivity were in the 1930s. Bell Telephone Laboratories conducted a series of experiments in communication. In the 1940s, diagnostic and aural rehabilitation were added. It became clear that determining threshold in dB SPL (sound pressure level) could be difficult and confusing. By the fact that hearing sensitivity varies across frequency, normal hearing, and subsequently hearing loss, would need to be defined differently at each frequency.

Over the years various organizations consolidated a large amount of research on the hearing threshold at many frequencies for young adults with no auditory pathology. Then, they took the mean threshold of this huge number of subjects and made it, as a reference, the "zero" decibel hearing level (dB HL). The organization who currently stipulates standards for audiometric assessment is the American National Standards Institute (ANSI).

The sensitivity to sound is one of the best ways to describe hearing ability. At the same time, one of the best ways to describe hearing disorder is by measuring the reduction in sensitivity to sounds. Hearing sensitivity is usually defined by a threshold of audibility of a sound that is individual to each subject. Specific measurements are needed to determine the just barely audible intensity of a tone or a word. That level is considered the threshold of audibility of the signal and is an accepted way of describing the sensitivity of hearing.

There are two main methods to measure the auditory threshold:

Objective Audiometry

These tests are based on recording the electrical activity of the different parts of the auditory pathway. This type of tests does not require the patient's participation or any verbal answers. Examples of different methods of objective audiometry are the Measurement of Impedance, the Measurement of the Otoacoustic Emissions (OAE) and the Measurement of the Auditory-Evoked Potential (AEP).

Subjective Audiometry

Unlike objective audiometry, subjective methods require the patient's cooperation and participation. These tests usually provide reliable quantitative measures on a subject's hearing function, if the subject understands the test procedure and is cooperative. A number of different subjective audiometry are showing as follows:

Pure-tone Audiometry

It is performed to obtain the auditory threshold. This threshold registers the minimum intensity of hearing in both ears for different audible frequencies. The stimuli are a number of pure sinus tones between 100 and 10000 Hz, presented monaurally to the patients. The pure-tone audiometry can determine whether there is conductive and sensorineural hearing loss.

Békésy Audiometry

Is an automated assessment in which the patient controls the attenuation of the signal. By pushing a button, the patient increases the intensity of the signal until it is audible. The listener then releases the button until the signal is inaudible, presses it until it is audible again, and so on. These responses are displayed on a screen and the threshold is calculated as the midpoint of the responses between audible and inaudible. While the tracking occurs, the frequency of the signal is slowly swept from low to high, so that an audiogram is measured across the frequency range [15]. This type of audiometry is, however, rarely used in diagnostics.

Speech Audiometry

Refers to procedures that use speech stimuli to evaluate auditory function. Speech audiometry involves the assessment of sensitivity as well as the clarity at which speech is heard. Several tests have been developed over the years. Most use single-syllable words in lists of 25 or 50 words. Lists are usually developed to resemble the phonetic content of speech in a particular language. Word lists are presented to patients, who are instructed to repeat the words. Speech perception is expressed as a percentage of correct identification of words presented. Speech audiometry can tell, in a more realistic manner than with pure sinus tones, how an auditory disorder might impact communication in daily living.

Speech audiometry measurements contribute in a number of important ways, including measurement of speech threshold, cross-check of pure-tone sensitivity, quantification of speech-perception ability, assistance in differential diagnosis, assessment of auditory processing ability, and estimation of communicative function [12].

The most common German speech audiometry, used by the majority of hospitals, medical practitioners, and hearing-aid dispensers, is the Freiburg speech test (FST) [99]. The FST is a standard test in hearing diagnostics and in the validation of hearing aid fittings. This test employs the use of phonetically balanced lists of monosyllabic words with the aim of determining the percentage of correctly repeated words at different sound intensity levels. FST consists of 20 lists with 20 monosyllables. Among the most important criticisms about this test are the differences between the test lists, the limited number of available lists, the use of outdated words, and the lack of the possibility to determine speech intelligibility in noise [125].

Tuning Fork Tests

It produces a sustained pure-tone that decays in level over time. Unlike an audiometer, tuning forks cannot present a calibrated signal level to a listener's ear. The two best-known tuning fork tests are the Weber and Rinne. For the Weber test, a subject judges whether the sound is perceived in one or both ears when the tuning fork is placed on the forehead. For the Rinne test, the listener judges whether the sound is louder when presented by air conduction or by bone conduction. Tuning fork tests provide qualitative information that can help to determine whether a hearing loss is conductive or sensorineural [62, 211].

Recruitment

An injury in the external hair cells means that weak signals are not perceived because they are not amplified. However, intense signals that directly impact the inner hair cells are normally perceived. This results in an abnormal perception of loudness known as recruitment. Recruitment is, therefore, a manifestation of cochlear injury in the external hair cells. Clinically this means that if recruitment is present, then the site of the disorder is cochlear [12].

Speech-in-noise Perception

The most common complaint expressed by adults with hearing loss is the inability to understand speech in an environment with background noise, due to the speech perception in a noisy environment is much more demanding than speech perception in silence. Audiologists use speech-in-noise tests for quantifying the signal to noise ratio (SNR) needed by the listener to understand speech in noise. There are several numbers of speech-in-noise test that have been developed over the years (see chapter 3). Speech-in-noise test allows the patient to understand the degree of communication difficulty they experience in noisy environments. The information provided by the speech-in-noise test allows a selection of the most appropriate amplification strategy as well as predicting the degree of improvement with the use of hearing aids.

Several speech-in-noise tests, in the German language, have been developed, such as:

The Basel sentence test [236]: It can be used to assess the degree to which the listener can make use of the contextual information in understanding the keywords (specific words that must be identified) of the speech material. Speech materials with high and low predictability are presented, and the background noise is an unintelligible babble-noise whose level is raised for the last word.

The Hochmair-Schulz-Moser sentence test (HSM) [107]: Consists of 600 everyday sentences arranged in 30 lists of 20 sentences. Additionally, there are six interrogative sentences in each list. The lists are prepared with 5 different levels of noise: without noise, SNR (in decibel HL) >15 dB, SNR > 10 dB, SNR > 5 dB and SNR > 0 dB. The sentences are played back in one ear, while in the other ear a CCITT noise (Committee Communication International Telephone and Telegram) [228] is being played as a masker. The level of the noise can be varied from list to list. The patient's task is to repeat the sentences that they heard. It was developed with the desire to evaluate speech perception of cochlear implant (CI) users [151].

The Göttingen sentence test [138]: Consists of 20 lists of 10 everyday sentences (5-8 words) spoken by an untrained male speaker with a speech-shaped noise as masker distractor. The Göttingen test can be used to measure speech performance at fixed SNRs or adaptively determine the speech reception threshold (SRT) (see chapter 3). Thus, the test is suitable for moderately hearing impaired and it is used in research and by advanced audiological centers. The downside of the Göttingen sentence test is its undesired learning effect [160].

The Oldenburg sentence test (OLSA) [238, 239, 240, 241]: It was developed and evaluated for testing speech intelligibility in noise and is also applicable to quiet conditions. The OLSA can determine the SNR where 50 % of words is understood (or speech reception threshold, see chapter 3) using an adaptive procedure or at fixed SNRs. The background noise is speech-shaped noise that is presented from a different loudspeaker to the one presenting the speech material. The test consists of 10 lists of 10 sentences with a fixed structure (5 words), combined to lists of 20 or 30 sentences. The speech material was recorded by an untrained male speaker. OLSA is suitable also for severely hearing impaired and cochlear-implant subjects [242].

Until today, the best single indicator of hearing loss and the prognosis for successful use of a hearing-device is the pure-tone audiogram. This audiogram, it has become the cornerstone of audiologic assessment and the generic indicator of what is perceived to be an individual's hearing ability [12, 226]. The pure-tone audiogram can be used to make judgments about several issues, such as separating normal from abnormal sensitivity. Despite many years of studies, there is no universally accepted criterion of what is normal hearing. This is partly because of all the individual aspects related to the audiometry and the different opinions about what level of hearing represent the onset of difficulty in day-to-day life. Despite all the discussions, the most used threshold definition is 20 dB HL as a normal cutoff.

The auditory sensation area, shown in Figure 2.1, is defined between two thresholds that are frequency dependent: hearing (the minimum level at which a sound can be detected) and pain (the level at which the sound becomes painful). The auditory area in humans is from 0 to 140 dB in a frequency range from 20 Hz to 20 kHz.

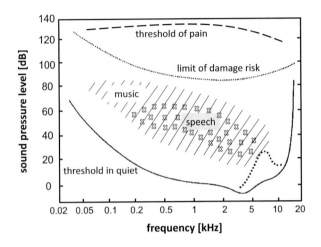

Figure 2.1: Auditory Sensation Area. Adopted from Zwicker and Fastl [267].

Despite all the benefits, the pure-tone audiometry gives only a cursory idea of the degree of difficulty in spoken communication caused by hearing loss because it does not take into account complex real-life acoustic scenarios as moving sounds sources.

2.2 Spatial Hearing

The sounds originate at a particular place in the space and its location could be important information, for example, for visual attention in cases of danger. Also, the position of sound sources can be used to separate between several sounds arising from different locations and to help in the attention to one specific sound source originated from a particular location.

The auditory system has just two peripheral spatial channels: the two ears. However, they can give us quite accurate information about sounds in the space. Listening with two ears is defined as binaural hearing; at the same time, monaural refers to the characteristics of one ear.

9

The human auditory system can be divided anatomically roughly into four subsystems: the outer ear, the middle ear, the inner ear, and the central auditory system.

The outer ear represents the most external portion of the auditory system and consists of the pinna, the outer ear canal, and the eardrum:

Pinna

It is composed of soft tissue and elastic cartilage. The surface is uneven and contains pits, depressions, ridges, and grooves (see Figure 2.2). It is known that for some species, like felines, the pinna may help to collect sound energy into the ear canal but this function is limited in humans because it cannot be moved toward a sound independently of the head.

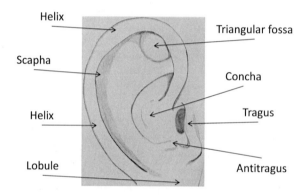

Figure 2.2: Anatomy of the pinna.

Ear canal

The ear canal is a canal in the temporal bone. Its size could vary from person to person depending on age and gender. Approximately, it is 25 to 35 mm in length with a diameter of 5 to 9 mm [145]. The end of the canal is limited by the tympanic membrane or eardrum. The ear canal protects the eardrum from the outside atmosphere, for that reason, usually is not perfectly straight (somewhat S shape) and cannot be seen by looking into the canal directly. Also, the canal acts as a resonator with a resonance frequency of 3 kHz approximately. This causes human hearing to be sensitive to that frequency.

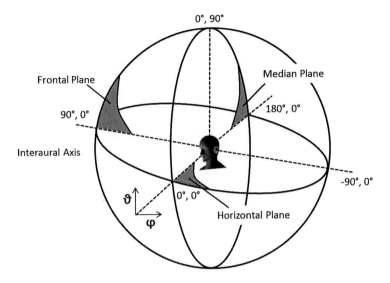

Figure 2.3: Head-related coordinate system. Adopted from [81].

Eardrum

The eardrum is a thin and very elastic membrane that separates the outer ear from the middle ear. The approximate dimensions are 90 mm^2 of surface with a thickness of 0.1 mm [12]. Sound waves impact the eardrum causing its vibration. This is a transference from airborne sound to structure-borne sound.

Figure 2.3 shows the Head-related coordinate system for sound direction, in which any direction relative to the head can be specified in terms of azimuth and elevation. The center of the system is the central point of the head, between the entrances of both ear canals. An interaural axis runs through the eardrums and the central point of the head. The horizontal plane is determined by the interaural axis and the front-back-link. The median plane cuts the head into two symmetrical parts thus any point on this plane is equidistant from the left and right ears. The third plane is the frontal plane which separates the head along the interaural axis.

Free-field HRTF =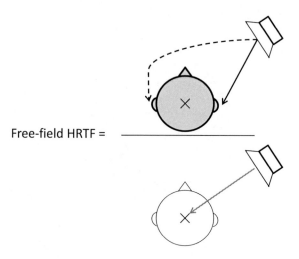

Figure 2.4: Descriptive definition of the free-field HRTF. Adopted from [81].

2.2.1 Head-related transfer function

The filtering of the sound source spectrum caused by the interactions of the sound waves with the head, shoulders, torso, and pinna prior to reaching the eardrum is represented by the head-related transfer function (HRTF) [24, 165]. Thus, an HRTF describes the filter process for any sound source spatially located in a room, for one of both ears. Accordingly, for one subject there is a dual channel band of HRTFs from all possible spatial directions and distances for both ears. The HRTF can be divided into two parts: dependent or independent of the direction of the sound. The dependent part is up to the entrance of the blocked aural canal because, depending on the location of the sound, the reflections in head, torso, and pinna will be different. Conversely, sound propagation, from the entrance of the aural canal to the eardrum is independent of the direction of the sound.

The usual definition of HRTF is the free-field transfer function. It describes the sound pressure measured at the entrance of the aural canal related to the sound pressure, measured with the same sound source, at the central point of the listener's head. The subject, however, is absent during the measurement (see Figure 2.4).

2.2.2 Binaural cues

In an every-day environment, several acoustical cues, arising from the environment itself (e.g. source distance, air propagation, reverberation etc.), as well as the physical composition (e.g. two ears spaced apart, notches and grooves of our pinna etc.), allow the perception of sound sources. Thus, listening with two ears provides access to binaural cues due to the difference in the travel path of a sound towards the two ears [2].

However, in a virtual environment, these cues must be simulated in order to reproduce (as closely as possible) the cues available under natural listening conditions. Given the fact that we have two separated ears, the different path lengths for the same sound to travel to the two ears result in the interaural time difference (ITD), whereas the intensity attenuation due to reflections and refraction around head and torso creates the interaural level difference (ILD) which is especially frequency dependent. ITD and ILD cues are known as binaural cues since they result from a comparison of the signals received at each ear.

Figure 2.5 shows three different sound source positions with different ITDs. The incident sound from the right $(90°, 0°)$ causes the largest time difference (ITD $\approx$ 690 μs). The sound from the front of the listener $(0°, 0°)$ reaches both ears at the same time, hence, there is no time difference (ITD = 0 μs).

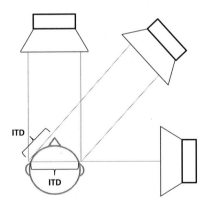

Figure 2.5: Descriptive scheme of three sound source positions with different ITDs. Adopted from [81].

Low frequencies ($<$ 500 Hz) are not influenced by the head and torso due to their long wavelengths. However, for frequencies higher than approximately 1500 Hz, where the wavelengths are smaller than the head, the wavelengths are too small to bend around the head and therefore are blocked by the head (e.g. "shadowed" by the head). As a result, the energy of the sound reaching the contralateral ear decreases (see Figure 2.6 and Figure 2.7). This difference, due to ILD, can reach up to 20 dB.

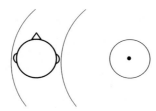

Figure 2.6: Descriptive scheme of the frequency dependence of ILD with low frequencies. Adopted from [81].

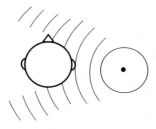

Figure 2.7: Descriptive scheme of the frequency dependence of ILD with high frequencies. Adopted from [81].

It is important to emphasize that the ILD only describes the spectral differences between both ears. All other spectral influences are called monaural cues (see subsection 2.2.3).

An important case to consider is when ITDs and ILDs are identical, respectively, for different source positions. All positions, for which this assumption is true, lie on a cone's surface. The cones are always positioned around the interaural axis and they end at the center of the head. Figure 2.8 shows a cone with five positions where the localization will be ambiguous. Positions on the cone of confusion are more difficult to differentiate with regard to those with different ITDs and ILDs.

A singular case of cone of confusion is a sound source directly in front or in the back of a listener (see Figure 2.9). In such a situation, both ITD and ILD will be negligible and the listener will not be able to determine whether the sound source is directly in front or in back of them, based on ITD and ILD cues only. This ambiguous case is known as the front-back confusion. However, it is possible to make use of the pinna cues (see subsection 2.2.3) to distinguish between back and front directions.

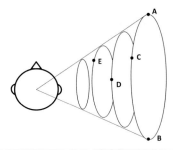

Figure 2.8: Descriptive scheme of five sound sources positions in the cone of confusion with identical ITD and ILD. Adopted from [81].

Figure 2.9: Sound source positions where ITD and ILD are zero, provoking confusion on the real localization of the source. Adopted from [81].

In addition to ITD and ILD cues, it is possible to make use of the asymmetry of the auditory system, through the right-ear advantage (REA). In general, ear advantage is defined as the relatively better performance of one ear over the other ear on listening tasks. A slight right-ear advantage for dichotic speech stimuli has been reported in normal healthy adults [17, 52, 106]. Doreen Kimura was the first to apply the dichotic listening method to neuropsychology and she discovered the right-ear advantage for linguistic stimuli, which is associated empirically with left-hemisphere language representation [129, 130].

2.2.3 Monaural cues

Although we have two ears, a large amount of information about sound source location can be obtained by listening through just one ear. The primary monaural cue is called a spectral cue because the information is contained in the distribution differences (or spectrum) of the frequencies that reach the ear from different locations. These differences are caused by the fact that before the sound stimulus enters the auditory canal, it is reflected from the head and within the various folds of the pinna. Because each pinna contains unique ridges and cavities, the pinna imposes a sort of directional signature on the spectrum of the sound that can be recognized by the auditory system and used as a monaural cue.

There is also a shadowing effect of the pinna for sounds behind the head, which have to diffract around the pinna to reach the ear canal. The shadowing will tend to attenuate high-frequency sounds coming from the rear and may help to resolve front–back ambiguities.

On the whole, each type of cue works best for different frequencies and different coordinates. Binaural cues (ITDs and ILDs) work for judging azimuth location, being ITD best for low frequencies and ILD for high frequencies. Monaural cues work best for judging elevation, especially at higher frequencies. These cues work together to help in sound perception tasks [91]. In real listening situations, it is also possible to move our heads, which provides additional ITD, ILD, and spectral information (for more detail, see subsection 2.2.6).

2.2.4 Binaural sluggishness

A number of investigators have studied the ability of subjects to follow changes in the location of stimuli over time [66, 69, 137]. Most of these studies have shown that only rather slow changes in location can be followed. This phenomenon has been described as "binaural sluggishness".

Perrott and Musicant [198] and Grantham [94] measured the minimum audible movement angle (MAMA), defined as the angle through which a sound source has to move for it to be distinguished from a stationary source. For low velocities, as 15°/s, the MAMA is about 5°, but when the rate of movement increases, the MAMA increases progressively up to about 20° for a velocity of 90°/s. Thus, the binaural system is relatively insensitive to movements at high rates.

Grantham and Wightman [95] measured the ability to follow movements of a noise that was lowpass filtered at 3000 Hz. Their results indicated that slow movements can be well followed, but rapid movements are more difficult to follow. The sensitivity to changes in binaural cues has also been determined by measuring masking level differences. Grantham and Wightman [96] measured thresholds for detecting a brief tone that was phase inverted at one ear in relation to the other. Again, their results indicated that the binaural system is slow in its response to changes in interaural stimulation.

In summary, research to date indicates that the binaural system often has a slow response to changes in interaural time, intensity or correlation.

2.2.5 Spatial unmasking

Masking is a phenomenon present in our daily lives. When the ears are exposed to two or more sounds at the same time, there is a possibility that one of them may "cover up" the other. A simple example is trying to have a conversation at a party. In general, the music is so loud that it is difficult or even impossible to understand what the other person is saying because the speech is masked by the music.

Many studies have been done about masking effect [50, 190, 191]. In the temporal domain, there are different types of masking. The most simple to understand is the simultaneous masking as in the previous example. However, there is also masking when a soft tone is close in time to a higher amplitude tone. According to the temporary position of the tone and the masker there are:

Forward masking
The tone of highest amplitude comes before the lowest amplitude tone, keeping on in that way masked.

Backward masking

The tone of lowest amplitude comes before the highest amplitude tone, keeping on anyway masked by the highest tone.

When faced with the hearing of speech, it is possible to distinguish two types of masking that can interfere with the speech signal:

Energetic masking

It occurs when the neural activity evoked by the speech plus masker is very similar to that evoked by the masker alone.

Informational masking

It occurs when the speech is confused with the masker in some way. When the speech of two talkers is mixed, there is often relatively little overlap between the spectrotemporal regions, dominated by one talker in some regions and by the second talker in other regions. In this situation, the ability to identify the speech of one talker is limited by informational masking, rather than by energetic masking [40]. The problem for the listener is to decide which spectrotemporal regions emanated from one talker and which are emanated from the other.

The auditory system is capable of performing spatial unmasking and this ability relies heavily on the magnitude of the available binaural cues. The contributions of ITD and ILD in unmasking tasks have been largely studied (e.g. [37, 67, 104, 149, 213]). For instance, Lavandier and Culling [149] explained that the spatial unmasking associated with two sound sources (target and masker) spatially separated on the horizontal plane, arises from two cues: head shadow and binaural interaction. When the masker moves around a listener's head, its sound level is reduced at the contralateral ear in the shadow region of the head, providing a higher SNR compared to the ipsilateral ear. In that way, the "head shadow" effect is the difference in SNR of each ear in regard to the target (also known as better-ear-listening) and it is very helpful in speech-in-noise perception (see Figure 2.10). Binaural interaction relies on different ITDs and ILDs resulting from the spatial separation of the target and distractor masker, which provides differences in the binaural cues to facilitate the segregation of the two sound sources. The contribution of ITD alone without the presence of ILD is as much as 5 dB release from masking for a $0°/90°$ target-masker configuration on the horizontal plane, but only values between 2.1 and 3.4 dB in the overall release from masking in natural listening situations where both ITD and ILD are present

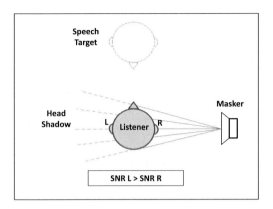

Figure 2.10: Schematic representation of the "Head Shadow" effect. L represents the left ear and R the right ear. The level ratio between the target signal regardless of the masker signal in the left ear (SNR L) is larger than for the right ear (SNR R). For that reason, in this example, the left ear has an advantage in speech-in-noise perception.

[37]. Most of these studies utilized stationary sound sources in the target and maskers, in which binaural cues remained unchanged (often assuming listener head movement was negligible). In the current thesis, the effect of a transient change in binaural cues, such as from a moving sound source (see chapter 6) or listener head movements (see chapter 8) are presented.

Past work has shown that spatial separation of a speech and a noise can provide a significant listening advantage in multisource environments for a variety of tasks. When the masking sources are primarily energetic, the listening advantage may result from attending to the ear with the more favorable signal-to-noise ratio (the "better ear advantage"), or from the binaural analysis. The principal binaural cues that afford a listening advantage due to the spatial separation of the sources arise from differences in interaural time-of-arrival at the two ears and the frequency-dependent differences in level at the two ears [124].

2.2.6 Head movements

In any normal listening environment, we are not immobile but we are free to move. In particular, it is possible to move the heads, from side-to-side, up and down or in any other way. A theoretical model proposed by Lambert [144] describes the mathematical mechanism in determining source location with head movement. It was later shown that non-human mammals such as cats [234] and monkeys [207] achieved better sound localization when their heads are unrestrained during the experiment. For human sound localization, several studies also showed the benefit of head movement which allows a solution of auditory ambiguities from front-back confusion using real sound sources and in a virtual acoustic environment [14, 118, 123, 253].

Head movements provoke changes of position between the sound source and the listener, leading to changes in the ITD and ILD cues. It is possible to integrate these changes as they occur over time in order to resolve ambiguous sound localization situations.

Consider a sound source directly in front of a listener. In such a situation, both ITD and ILD will be negligible and it will be very difficult for the listener to determine whether the sound is directly in front or behind. But if the listener rotates his/her head (e.g. to the left) by a certain amount of degrees, the ears are moved from their initial position to some new position. Although the sound source has not moved from its initial location, relative to the listener, the sound source position has changed. As a result, there are two observations of ITDs and ILDs that together allow disambiguating the cone of confusion and a more accurate localization is possible.

As well as the head movement seems to improve localization of both the target and masker sources, it might also improve target intelligibility in a speech perception task, but evidence from recent studies did not bring sufficient support for the use of such strategy [33, 123]. To study the strategy of using head movement more rigorously with better control of confounding factors, Grange and Culling [93] presented various configurations of target-masker spatially separated using loudspeakers on an array as real sources. These spatial configurations were made to induce head turns among normal-hearing listeners to maximize spatial release from masking (see section 3.3), but head tracking data of free head movement suggested that listeners did not pursue such a strategy. Since distractors in real-world are not always stationary, such as a moving talker or a passing vehicle, this thesis also examines head movements and speech perception on a moving masker test (see chapter 8).

2.3 Binaural Reproduction Technique

Since the creation of the telephone in the late 1800s and the radio in the 1900s, there have been many developments and improvements in the technologies for presenting sounds to a listener. The exact reproduction of the original sound field, including all spatial cues (e.g. binaural and monaural cues), is certainly the goal of most sound reproduction technologies.

The first recording techniques involved recording a sound field (e.g. a concert) using one or more microphones and then playing back using one or more loudspeakers or a pair of headphones. When the assessment of speech perception in noise started to play an important role in different areas as telecommunication, room acoustics, audiology, and evaluation of hearing aids, more complex recording techniques were needed to reproduce a more realistic environment.

Many studies have been conducted to assess speech-in-noise perception under different conditions and, to achieve this objective, several techniques have been developed such as stereophonic [36, 201], ambisonics [87, 186] and binaural reproduction [37, 38, 74, 105, 180, 181]. The advantage of binaural reproduction techniques are the low computational demand and the small amount of equipment needed. In the case of clinical applications, techniques with a large number of loudspeakers, such as stereophonic and ambisonics, may not be applicable because audiologists usually count with a small sound-treated cabinet to perform the listening tests.

2.3.1 Binaural recording

The binaural reproduction technique relies on the fact that all spatial sound information perceived by humans is extracted exclusively from the sound signal (pressure and phase) that reaches each of the eardrums. If these signals are recorded in the ears of a listener and reproduced exactly as they were, the listener may use any of the naturally available cues in order to perceive the sound as coming from the desired position. In this way, it is assumed that the complete auditory experience is reproduced, including timbre and spatial aspects. [170].

The recording may be made with small microphones placed in the ear canals of a listener, but normally an artificial head is used. The artificial head or dummy-head, has usually the shape of an average human head, including the nose, pinna and ear canals, and sometimes the head is even attached to a torso [101, 116, 181, 193, 214].

The binaural signals can either be directly recorded in a dummy-head or they can be synthesized. In the binaural synthesis, each sound source must be filtered with a head-related transfer function (HRTF) (see subsection 2.2.1). HRTFs are, hence, listener-dependent [171, 246], therefore, when using binaural technology to create an authentic spatial sound scene, the use of individual HRTFs could be more accurate. A comparative analysis between individual and non-individual HRTFs is presented in section 8.4.

2.3.2 Binaural synthesis

Rather than recording the signal at the ears for a specific listening situation as done with binaural recordings, binaural synthesis imitates the binaural recording process by convolving a monaural sound source with a pair of left and right ear HRTFs corresponding to the desired position, typically measured at an anechoic room. The signals delivered to the left and right ears are obtained using a filtering operation through the process of convolution by filtering the monaural sound with the coefficients corresponding to the measured left and right ear HRTF response, respectively. When the filtered sound is presented to the listener, it will give them the impression of a sound source at the desired position.

The playback is normally done with headphones since this method ensures that the sound that reaches one ear is only reproduced in that ear (separated channels). Reproduction through loudspeakers would introduce an unwanted crosstalk since sound from each of the loudspeakers would be heard with both ears. For these cases, it is possible to filter the binaural signal using a Crosstalk Cancellation Filter (CTC) network to minimize the acoustic crosstalk between the ears [165]. A binaural synthesis based on measured HRTF is often for a static receiver-source situation. However it can also be conducted in real-time using a head tracking system and updating the HRTF based on the current position of the listener in the scene.

Given all its advantages, this thesis was developed using a binaural reproduction technique.

2.3.3 Headphone transfer function

Since in this thesis a pair of headphone is used to playback the stimuli, the acoustic influence of these headphones has to be considered. With headphones, the sound is directly played back over the external ear into the ear canal to the eardrum. Consequently, if headphones are used for the binaural playback with free-field HRTFs, which already provide information about the human body, the influence of the headphones, external ear, and ear canal has to be compensated [251].

Although the ear canal is taken into account in both, the free-field and headphone listening condition, the transfer function of the ear canal is position-dependent [27]. Therefore, the measurement position of the microphone inside the ear canal has to be considered [182]. In particular, waves that enter the ear canal are reflected in the canal so that standing waves occur above 3 kHz. Two measurement positions are very common: At the ear canal entrance (blocked-ear) and in front of the eardrum (open-ear). Although the position at the eardrum is more clearly defined than the entrance of the ear canal, it is very sensitive to noise. Therefore, the blocked-ear method with a microphone at the ear canal entrance is preferred more often; however, the entrance of the ear canal cannot be determined precisely but it can be estimated roughly 7mm in front of the eardrum for frequencies below 6 kHz. The headphone transfer function (HpTF) has to be measured in the same position as the HRTF. Consequently, for the free-field reproduction, the playback signal has to be multiplied by the HRTF and divided by the inverse of the HpTF in the frequency-domain [172, 252]. There are different methods to determine the inverse of the HpTF which equalize the headphones and the transfer path of an emitted wave to the eardrum. Masiero and Fels [166] proposes to smooth the spectrum of the HpTF by adding twice the standard deviation of eight repeated measurements to their average to provide a smooth and robust equalization with respect to outliers. In this work, the proposed method of Masiero and Fels [166] was used.

2.3.4 Individual head-related transfer function

The use of HRTFs in sound reproduction applications describes the direction-dependent influence of torso, head and pinna on the sound field, and is therefore highly dependent on individual anatomical characteristics. If a virtual acoustic scene, which was convolved with a dummy-head HRTF, is playback to a subject; there will be some mismatches between the synthetic HRTF and the own HRTF of the subject. It was established that the usage of mismatched HRTF data-sets can result on unwanted artifacts in coloration and localization [27]. To avoid this mismatch, HRTFs measured from the individual listener are needed.

Little is yet known about the influence of individual HRTF on speech perception in noise tests, thus a comparative analysis between individual and non-individual HRTFs is presented in section 8.4. The setup used in this work to measure the individual HRTFs was designed by the Institute of Technical Acoustics (ITA) at RWTH Aachen University (Figure 2.11), providing a high-resolution in a very short time period [167, 208]. To reduce the acoustic influence of the measurement aperture, the setup had to be built in a filigree manner. In the vertically aligned

Figure 2.11: Measurement setup of the arc and an artificial head on a rotating turn table at ITA, RWTH Aachen University.

continuous arc, which provided the cavity for the loudspeakers, 64 loudspeakers were installed. They were placed in polar direction in a resolution of 2.5° on a semi-circle starting at 1.55° and ending at 160°. Measurements took place in a semi-anechoic chamber with a stone floor which reflects incident waves.

Prior to the start of the measurement, two Sennheiser KE3 microphones, supported by a dome which blocks the ear [182], were positioned at the beginning of the ear canal. Subsequently, the subjects were aligned using a cross-laser. Additionally, a neck support minimized the head-movements during the measurement. For the spherical sampling, the turntable was moved in discrete steps and performed a full rotation of 360° (for more details see [27] and [208]).

3

Review of Speech-in-Noise Perception

Auditory assessment using speech stimuli has a long history in the evaluation of hearing. As early as 1800, studies of hearing sensitivity were created with the purpose of evaluating the listener's capacity to perceive different types of speech sounds. The speech stimuli consisted of: (1) vowels; (2) consonants; (3) combinations of vowels and consonants; or (4) monosyllables. These studies continued through the 19th century, thus by the mid-1920s the first speech audiometer, the Western Electric 4 A, which incorporated a phonograph with recorded digit speech stimuli, was employed in large-scale hearing screenings [80]. Hearing and understanding speech is of utmost importance in our lives. For children it is essential to the development of oral language and for adults it is fundamental to participate in the numerous communicative interactions activities of daily living. The measurement of understanding and sensitivity form the basis of speech audiometry, which refers to procedures that use speech stimuli to assess auditory functions [140]. Speech audiometry has involved the assessment of (1) audibility component (i.e., loss of sensitivity), and (2) distortion component [204]. The audibility component is quantified through assessment of speech recognition abilities in quiet. The distortion component is a reduction in the ability to understand speech in the presence of background noise, regardless the level of presentation. The quantification of the distortion component typically involves the evaluation of the percentage of correctly identified words using the word recognition score (WRS), that is based on the recognition of monosyllabic words. More recently, the speech reception threshold (SRT) has been recommended instead of the traditional WRS [128, 256]. Plomp [204] defined SRT as the signal-to-noise ratio (SNR) at which the 50 % of the speech material is correctly recognized. In this case, the speech material could be words (without restriction in the number of syllables), sentences, or continuous speech; thereby, the SRT can assess a more "realistic" listening condition for everyday communication.

In everyday acoustic environments, listeners often need to resolve speech targets in mixed streams of distracting noises. Furthermore, the most common complaint expressed by adults with hearing impairments is the inability to understand a speaker when listening in an environment with background noise. For that reason, several researchers have been supporting the relevance of measuring speech perception with a noise background [13, 47, 73, 98, 126, 185, 258]. Word identification tasks are typically used for measuring speech-in-noise perception. The task requires listeners to write down what the speech target is saying. An open response format is most commonly used with the listeners instructed to write down what they think the speech target is saying. In some cases closed-set tasks are used, where listeners are given a range of multiple-choice alternatives from which to select their responses. Scoring procedures vary but typically involve sentences being scored on the number of keywords (specific words that must be identified) correctly identified or by the total number of words correctly identified.

3.1 Speech Reception Threshold

The most relevant factors that influence speech-in-noise perception are: the type of target speech, spectral differences between target and masker, spatial configuration of the sound sources, fluctuations in level of the masker, the room acoustic conditions, and hearing impairment of the listener. The effects of these factors often are quantified as shifts of the SRT [35]. The SRT has been defined as the minimum intensity at which spoken language can be understood (50 % of recognition) under the presence of masking noises. In other words, the SRT calculation is a psychometric function that measure changes in a dependent variable (y-axis; e.g., number or percent correct, which is a psychological variable) based on changes of an independent variable (x-axis; e.g., presentation level in SNR, which is a physical variable) [86]. Figure 3.1 show a graphic display of a psychometric function. In this example, the speech perception is evaluated for 0, -4, -8, -12, -16, -20, -24, and -28 dB SNRs (black dots). As can be seen, the percent of correctly identified words is low when the SNR is low, and as the SNR is increased, the percent of correctly identified words increases. The dashed line in Figure 3.1 highlights the 50 % of correctly identified words and indicates that the SRT was around -14 dB SNR.

The slope of the function is also relevant when describing performance in terms of the psychometric function. The slope is typically calculated from the dynamic portion of the function that ranges between 20 % and 80 %. Scores below 20 % are often affected by floor effects because the difficulty of the task is too great to show subtle changes in performance, whereas scores above 80 % are often affected by ceiling effects because the difficulty of the task is too easy to be sensitive to changes in the performance. When selecting speech target material, stimuli that

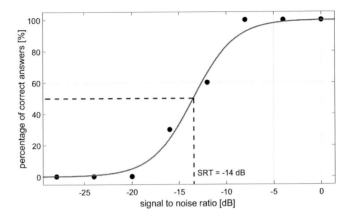

Figure 3.1: Psychometric functions of word recognition performance measured in percent correct (y-axis) for a listener as a function of presentation level (x-axis). The dashed line indicates the 50 % point. The SRT represent the level (SNR) at which the listener is able to recognize the 50 % of the words.

produce a steep function are the best choice, because suggests that the materials are homogeneous regarding to the task [259].

The first efforts to assess speech perception were performed in the 1930s by Bell Telephone Laboratories. Since then, many studies have been carried out with the intention of evaluating factors that affect our speech perception such as binaural hearing, the influence of interaural differences, reverberant conditions, the spatial location of sound sources, and different type of maskers [46, 84, 102, 134, 136, 152, 156, 206, 219, 233]. After all these studies, the authors agreed that creating reliable tests, by which the SRT can be determined, was necessary.

3.2 Reliable Speech-in-Noise Tests

Plomp and Mimpen [202] were one of the first proposing a reliable test to assess SRT. Even today, this procedure is one of the most used, therefore, a proper explanation is presented showing the necessary steps to obtain a reliable test:

The speech material
The speech material was sentences, due to represent conversational speech; nev-

ertheless, they must be short enough to be easily repeated.

Evaluation of the speech material

All sentences were evaluated by ten audiologists and speech therapists. After evaluation and equalization, 10 lists of 13 sentences were developed. The sentences were pronounced by a female speaker with a trained voice to avoid dialect influences.

Reproduction of stimuli

The reproduction of the sentences was binaurally through headphones at a level of about 50 dB (A).

Masker types

The spectrum of the noise masker was equal to the long-term average spectrum of speech. The advantage of adopting this masker spectrum is that the effect of accidental differences between the spectrum of the speaker and the noise is eliminated.

Adaptive procedure to calculate SRT

The main purpose of the test was to investigate the speech perception threshold under various conditions. The adaptive procedure to calculate the threshold is as follows: (a) the first sentence is given with increasing sound level until the listener can reproduce the sentence correctly; (b) the level is decreased by 2 dB and the second sentence is presented; (c) if the listener is able to correctly repeat the full sentence, the level is decreased by 2 dB, if he/she is not, the level increase by 2 dB; (d) repeat the previous step for all sentences (simple up-down procedure [153]); (e) the speech perception threshold is adopted from the average presentation level of the last eight sentences. Plomp defined it as the speech reception threshold (SRT) and represents the signal-to-noise ratio at which the listener correctly identify the 50 % of the speech material [204].

Reliable test

If the test is repeated with another list, the two SRTs values should be similar. The Figure 3.2 shows an example of a similar adaptive procedure to calculate the SRT. At the beginning, the step size is 8 dB until the first reversal is reached (5th trial), from which the step size change to 4 dB until the second reversal is reached (9th trial); from there on the step size change to 2 dB. The procedure finished when the listener reached the sixth reversal. The SRT is calculated by averaging the SNR level of the last four reversals: third reversal at the 11th trial (SNR = -20 dB), fourth reversal at the 12th trial (SNR = -18 dB), fifth reversal at the 14th trial (SNR = -22 dB), and sixth reversal at the 16th trial (SNR = -18 dB). The Figure 3.2 show that, at the end of the adaptive procedure, an SRT of -20 dB was achieved (red dash line).

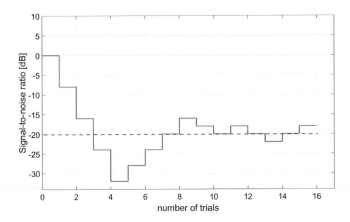

Figure 3.2: Changes in SNR during an adaptive procedure to track the SRT (SNR at 50% speech intelligibility).

After the test developed by Plomp and Mimpen, the speech perception in noise (SPIN) test was created [22, 174]. Some differences with the previous test were: (a) the speech material was sentences with low and high predictability; (b) the target word that must be repeated by the listeners (or keyword), was the last monosyllable of the sentence; (c) the masker was a multitalker 12-voice bubblenoise. For the authors, this test would provide a more useful index of "everyday speech reception" than word materials presented without contextual cues.

After those approaches, different studies were made and many tests were created to measure the SRT. With the purpose of assess soldiers hearing, the Speech Recognition in Noise Test (SPRINT) [56] was created, involving the presentation of 200 monosyllabic words in background noise (multi-talker babble). The SPRINT is administered by an audiometer through earphones with the stimuli delivered to both ears simultaneously. The monosyllabic words and the background babble are pre-recorded at the proper SNR on the same channel of the tape. The test score is simply the number of monosyllabic words correctly identified. The SPRINT was developed to assess if the soldiers were able to perform their duties, but it does not represent the needs of the general population in a common real-listening situation. The Speech in Noise (SIN) test [82, 127] was designed to evaluate speech understanding in noise (four-talker speech babble) for both soft (presented at 40 dB HL) and loud (presented at 70 dB HL) speech in a range of SNRs that involve easy to very difficult conditions. The test consists of nine blocks of 40 sentences each and the listener must repeat the last five words. The test is presented at 20, 15, 10, 5, and 0 dB SNRs. Sentences as target speech is assumed as more representative of a real-listening situation, but the test assess only one spatial configuration with the target at 0° and the masker at 90°. The Hearing in Noise Test (HINT) [178] uses sentences derived from the Bamford–Kowal–Bench (BKB) [16] sentences with semantic and syntactic context. The speech materials consist of 250 sentences presented in sets of 10 sentences. When the whole sentence is correctly identified, it counts as a correct answer. The masker used is speech-spectrum noise that is held constant in level, while the speech signal is varied to find the SRT. This test is largely used and it was translated into several languages. The Words-in-Noise (WIN) test [255, 257] consists of 70 monosyllabic words spoken by a female speaker and it provides a protocol to evaluate the abilities of listeners understanding speech in a speech-babble noise consisting of three females and three males talking simultaneously about different topics (not intelligible). The test is presented at seven SNRs from 24 to 0 dB, decreasing in steps of 4 dB, with a stopping rule of ten incorrect words at one level. The WIN was validated by other studies [168, 254], but the use of monosyllabic words could be not enough to represent real conversations. The Realistic Hearing in Noise Test Environment (R-HINT-E) [235] consists of 250 sentences from the HINT test, divided into 25 lists of 10 sentences each recorded by 6 males and 6 females for a total of 3000 sentences. Several configurations of target and masker locations are available for three different room acoustic conditions. The characterization of the room environment was made by impulse measures using loudspeakers at desired positions in the space (in azimuth 45°, 90°, 135°, 225°, 270° and 315°) and a dummy-head with two microphones in the ears recording the impulses. The test presents the option to assess several target/masker configurations, but the technique used does not permit a different assessment in another desired configuration. The digit-triplet speech-in-noise test [222, 223] consists of 5 lists of 24 digit-triplets randomly chosen from the set of 120 available digit-triplets. Each triplet contains three digits between 0 and 9 and the masker is a long-term average speech spectrum (LTASS) noise. The test uses an adaptive procedure to calculate the SRT. The Quick Speech-in-Noise (QuickSIN) test [128] is composed of the Institute of Electrical and Electronics Engineers (IEEE) sentences [210] that contains semantic content. The masker is a four-talker babble noise and the entire test is playback through headphones or loudspeakers. Each of the twelve lists contains six sentences with five keywords each. Despite delivering faster results than the SIN test, the sentences provide contextual cues that not all patients could understand. The Listening in Spatialized Noise (LISN) test [45] is an adaptive speech test that utilizes ten continuous discourses (stories written by a novelist) as the target stimulus and looped sentences as the masking noise. The stories are ranged in length from 3 minutes and 1 sec to 4 minutes and 42 secs. The test investigates the listener's ability to understand a story when a masking distractor is coming from different locations in a virtual auditory space and when the maskers are spoken by either the same speaker as the target or by different male and female speakers. The Speech Understanding in Noise (SUN) test [189] is based on the task of multiple-choice recognition of short meaningless stimuli in noise, specifically, vowel–consonant–vowel (VCV) stimuli (e.g., asa, apa, etc.). The background noise is a steady speech-shaped noise, i.e. a wide-band noise generated by filtering a steady-state unmodulated white noise with the reference long-term average speech spectrum for the specified language. Typically, 12–18 stimuli are used across the different languages and the test duration is below 1 min per ear. Thus, the SUN can be easily implemented and adapted to different languages, nevertheless, the meaningless stimuli do not represent real conversations.

In general, the differences lie on: the speech material, masker type, spatial and spectral configuration of sound sources, and the means by which the stimuli are presented to the listeners. The choice of the speech material is critical to assess speech-in-noise perception in more natural situations. Besides, the listener must be capable to understand the speech content correctly at a favorable SNR. Some listeners cannot understand the sentences correctly because of the severity of their hearing impairment; others cannot understand whole sentences or a complete story correctly because of limited linguistic skills. Therefore, the test must minimally depend on linguistic skills and should be feasible for listeners with hearing loss up to severe hearing impairments (desired also for CI users, and children). It seemed essential the use of simple familiar words instead of sentences, to reduce the effects of linguistic skills on the test result. One category of highly familiar words is comprised of digits. They are in the lists of the most frequently spoken words, they are known by children at a young age and they are typically among the first words that are learned in a second language [223]. For that reasons, in this thesis the digit-triplet test was selected as the speech material (see chapter 5).

3.3 Spatial Release From Masking

Is well known that listeners have a greater capacity to identify two (or more) sound sources as different if the sources are spatially separated than co-located [65, 105, 134]. This ability is helpful in many social situations where listeners are exposed to several sound sources at the same time such as a crowded street or a full restaurant. When speech and masker signals are playback at the same time, it is possible to gain an advantage in speech perception if the masker is located at a different position than co-located with the speech. This advantage is known as the spatial release from masking (SRM). It is known that SRM can be considered as having two components that can provide advantages in understanding the speech target. One advantage arises from improvements in signal-to-noise ratio (SNR) at the "better" ear, resulting from the head shadow effect, which is facilitated by the interaural level difference (ILD) that is particularly robust at high frequencies. The second component arises from the binaural advantage, which is strongly related to the interaural time difference (ITD) of the low-frequency content of the sound sources (see subsection 2.2.5).
The SRM is calculated as:

$$SRM = SRTcollocated - SRTseparated,$$

where $SRTcollocated$ is the SRT calculation in the case when target and masker are in the same spatial position (in general at $0°$ azimuth), and $SRTseparated$ is the SRT calculation in the case when target is at $0°$ and the masker is in a different spatial location (see Figure 3.3).
Many studies have been conducted to investigate SRM benefits. Bronkhorst [34] reported that there were differences on SRM depending on the method and speech material used during the measurement. Plomp [201] used a reverberation room with varying amounts of inserted sound-absorbing material to show that the SRM is largely abolished in the presence of reverberation, besides he showed that there were different SRM benefits if the masker is connected to speech or noise. Kidd et al. [124] concluded that SRM, using a masker with the same spectrum as the target, decrease from 7.9 dB in the very low reverberant condition to 2.0 dB in a high reverberant condition, this, when both sound sources were separated $90°$. Marrone et al. [164] showed that, when both sound sources were separated $90°$ in a reverberant environment, listeners with hearing loss seems not to be able to make use of the spatial separation benefits as well as the normal-hearing listeners.
Bronkhorst and Plomp [39] examined different quantities and spatial distributions

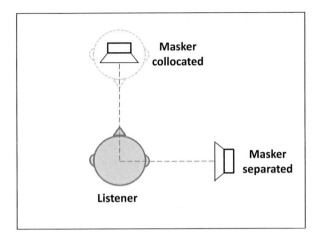

Figure 3.3: Example of masker configuration to calculate SRM.

of maskers. They found that when the number of maskers increases, the SRM decrease, furthermore when the maskers were in a symmetric array related the listener position, the SRM was lower than when maskers were in an asymmetric array.

Hawley et al. [104] examined different types of maskers under different spatial conditions for one, two or three maskers. They found that monaural spatial advantage disappear once multiple maskers are spatially distributed on both quadrants of the median plane, implying that head-shadow effect plays a minor role in listening situations when maskers are distributed in both quadrants of the median plane. On the other hand, the binaural advantage was robust in all spatial configurations: masker on both quadrants of the median plane or in only one.

Westermann and Buchholz [247] varied the maskers in distances of 0.5 m, 2 m, 5 m and 10 m, meanwhile, the target was fixed at 0.5 m. Their results showed an SRM of approximately 10 dB for speech masker at 10 m, but no spatial separation advantage for speech-modulated noise masker at 10 m.

Therefore, many factors contribute to an increase or diminish SRM including the measurement paradigm, room acoustic conditions, type of masker, the number of maskers in the scene, and its spatial distribution. However, while most current literature focuses on understanding the benefit of spatial separation with stationary maskers, little is yet known about the effect of spatial separation when the masker sound source is moving along a trajectory.

3.4 Predictive Auditory Processing Models

Due to the relevance of assessing speech-in-noise perception, many efforts have been made to predict speech intelligibility under the influence of noise, reverberation and hearing loss. The creation of predictive models to assess SRT and SRM may help in understanding the underlying mechanism of binaural hearing and may assist in the development and fitting of hearing aids [20]. Models of auditory processing may be roughly classified into biophysical, physiological, mathematical (or statistical), and perceptual, depending on which aspects of processing are considered [120].

Models of binaural speech intelligibility use peripheral preprocessing modeling outer/middle ear, basilar membrane, and haircells. The models convert the signals arriving at the ears into an internal representation where a diverse type of maskers, different spatial positions, psychophysical performance, and the influence of the room, could be predicted [20, 21, 57, 119, 120, 149, 132].

Numerous studies are concerned with measuring the SRM. A well-known SRM mathematical model was proposed by Bronkhorst [34], predicting SRM for any maskers configuration in the horizontal plane given a speech target located at $0°$ azimuth. The mathematical model is presented as follow:

$$SRM_{Bronkhorst} = C\left[\alpha\left(1 - \frac{1}{N}\sum_{i=1}^{N}cos(\theta_i)\right) + \beta\frac{1}{N}\left|\sum_{i=1}^{N}sin(\theta_i)\right|\right] \quad (1)$$

In this model, N is the number of maskers, θ_i is the angular separation of the i-th masker source with respect to the target, with $i = 1, \ldots, N$, given in degrees, and C represents an overall scaling coefficient that reflects differences among testing paradigms. The values of the regression coefficients α and β are 1.38 and 8.02, respectively. This model splits SRM into two additive components: "separation", i.e., angular separation of the masker from the target, and "asymmetry", i.e., degree of asymmetry of the masker configuration, expressed as:

$$SRM = SRM_{separation} + SRM_{asymmetry} \quad (2)$$

Jones and Litovsky [121] found significant differences between their results and the predictive model of Bronkhorst, in the first 45 azimuth angles of the $SRM_{separation}$ component. This fact made them formulate a revised model of

SRM, where the $SRM_{separation}$ component was characterized by a new mathematical function and regression coefficients as follow:

$$SRM_{J\&L} = D\left[\alpha\left(\frac{1}{N}\sum_{i=1}^{N}tanh(3\theta_i^*)\right) + \beta\frac{1}{N}\left|\sum_{i=1}^{N}sin(\theta_i)\right| + \right.$$

$$\left. \gamma\frac{1}{N}\sum_{i=1}^{N}\frac{1}{1 + e^{-0.5(|\theta_i|-110)}}\right] \quad (3)$$

$$With \quad \theta_i^* = \begin{cases} |\theta_i| & -90° \leq \theta_i \leq +90° \\ |\theta_i - 180| & +90° \leq \theta_i \leq +180° \\ |\theta_i + 180| & -180° \leq \theta_i \leq -90° \end{cases} \quad (4)$$

The scaling factor D serves a similar purpose as the overall scaling factor C in Bronkhorst's model. It describes magnitude differences of the SRM function, which depends on factors such as reverberation time, masker type, the measurement paradigm, number of maskers, target-masker similarity, and the relationship between masker angle(s) and SRM. N is the number of masker sources, with all maskers having the same long-term average level, θ_i is the azimuth angle of the i-th interfering source, with $i = 1,\dots, N$. A case differentiation for masker angles θ_i^* is defined and it lies between $0°$ and $90°$ depending on the respective masker angles θ_i. For noise masker, the model is defined for the entire horizontal plane, and the regression coefficients are $\alpha = 0.23$, $\beta = 0.75$ and $\gamma = 0.15$. In the case of speech maskers, where the model is defined for maskers in the frontal quadrant of the frontal plane only $(-90° to + 90°)$, the regression coefficients are $\alpha = 0.60$ and $\beta = 0.41$, with $\gamma = 0$.

Until now there is no information about an auditory processing model that could predict the influence of moving sound sources in SRM.

3.5 How Reverberation Affects Speech-in-Noise Perception

Reverberation is known to reduce the binaural differences generated by the masking sound thus affecting the SRM [64]. Besides, reverberation distorts the masking sound in such a way that temporal dips are filled in and distorts the target speech so that it becomes less intelligible [63, 111].

One of the most powerful strategies of the human auditory system to manage noise and reverberation is the spatial directivity, which is beneficial as long as interferers and reflections show different spatial patterns than the target speech. The auditory system has two related mechanisms to introduce directivity: the directivity of the ears, and binaural unmasking. The directivity of the ears is a consequence of the shape of the pinna together with the better-ear-listening (see subsection 2.2.5). This causes frequency dependent directivity patterns introducing ILDs which depend on azimuth and elevation of the sound source. Thus, the auditory system can use the ear with the better signal-to-noise ratio (SNR) while disregarding the ear with worse SNR. In many daily situations better-ear-listening is the most beneficial mechanism for speech perception in complex listening conditions. Another important ability of the auditory system is to use ITDs or interaural phase differences (IPD) for binaural unmasking [32]. Spatial unmasking depends on the azimuth separation of sound sources because the head shadow contribution is very dependent on the azimuths location of sound sources. The contribution of the binaural interaction proved to be relatively independent of the azimuth separation between sound sources, as long as they were not co-located. With increasing reverberation, the head shadow component progressively disappears [201], and spatial unmasking is reduced to its binaural interaction component. Consequently, it becomes independent of the amount of angular separation between sound sources [20, 201]. When the investigations only considered the spatial unmasking associated with binaural interaction, their results should not depend on the magnitude of the tested azimuth separations [148].

Plomp [201] investigated SRM measuring the intelligibility of a connected discourse as a function of both spatial separation and reverberation. The experiment was performed in a room having variable acoustic characteristics, with loudspeakers surrounding the listener. The SRM observed in the anechoic condition was greatly reduced in reverberant conditions. He found that a spatial separation of $90°$ improved masked threshold for a speech signal presented with a speech masker by about 5 dB in an anechoic condition and about 2 dB in a highly reverberant (2.3 s reverberation time) environment. The corresponding advantages using a speech-shaped noise masker were also about 5 dB in anechoic condition but less than 1 dB in the highly reverberant room. It was concluded that SRM was inversely related to reverberation time, suggesting that distortions of the spatial cues can provide reduced speech intelligibility due to reverberation.

Culling et al. [68] measured SRTs using speech as masker under headphone reproduction. Target and masker were assessed on two forms of intonations: monotone and intonated. They used a virtual room simulation that allowed positioning the sound sources at chosen positions, as well as varying the absorption coefficient of the room boundaries. Culling et al. compared SRTs for sentences in reverberant vs. anechoic conditions at different spatial locations.

They found a significant improvement due to spatial separation of sources for both intonated and monotone stimuli in the anechoic condition with thresholds of the monotone speech of 3–4 dB higher than for intonated speech. In the reverberant condition, thresholds increased for both types of speech, especially for the spatially separated presentation, causing the spatial separation advantage to be reduced to less than 1 dB. The spatial unmasking observed in the anechoic condition was abolished in reverberation.

Beutelmann and Brand [20] measured SRTs with a noise masker, creating their stimuli from binaural impulse responses measured in three different rooms: an anechoic room, a small office and a large cafeteria. Again, the SRM was reduced in the office and the cafeteria compared to the one obtained in the anechoic room. Zurek et al. [266] found that when the masker was farther away from the listener than the target, it increased the amount of the detrimental effect of the reverberation, leading to higher masker thresholds.

Kidd et al. [124] measured SRM with noise and speech as maskers, in three different reverberant conditions (FOAM, BARE and PLEX, from low to high reverberation). For the noise masker condition, the mean SRM decreased from 7.9 dB in the FOAM condition to 2.0 dB in the PLEX condition. The magnitude of the SRM depends significantly on the type of the masker.

A binaural room impulse response (BRIR) represents the acoustical transformation for a sound source in a room as measured at each of the listener's ears, and therefore it is influenced by factors such as the acoustic characteristics of the room, the position and orientation of the source, the position and orientation of the listener and the locations of other sound reflective objects in the room such as furniture. The sound reaching a listener's ear in a reverberant room consists of three main components: the direct sound (DS), the early reflections (ERs), and late reverberation (LR). The DS is the sound that reaches the listener before interacting with any surfaces and thus it is not influenced by the acoustic characteristics of the listening environment. ERs are defined as reflections that arrive immediately (<50 ms) after the direct speech. Perceptually, studies have shown that strong ERs improve speech intelligibility by increasing the loudness of direct sound [224]. On the contrary, LR (that arrives after early reflections and is made up of a very large number of delayed and attenuated copies of the original signal with different amounts of amplitude and phase distortions) is not perceptually integrated with the DS and it interferes with the intelligibility [225].

3.6 Studies on Moving Sound Sources

In real-life listening situations, we are confronted with multiple sound sources, either stationary or moving, that disturb our speech perception. When the target and masker streams are spatially separated, the differences in binaural cues available to the listener became important resources for a better speech perception (see [34] for a review). In natural acoustic scenes, conversations may become very difficult to understand with masking noises that have movements in space. Due to technical limitations in creating moving sound sources in listening experiments, little is yet known about the effect on intelligibility when the sound sources are moving.

The effect of SRT has been studied by including single or multiple masker sound sources, which were typically located at static positions [188, 260]. Studies focusing on moving sound sources have investigated perception and sensitivity of auditory motion and velocity [48, 49, 159, 161], others have evaluated speech recognition but when target and maskers suddenly changed their spatial positions during the stimulus reproduction [7, 42]. Only a few studies with sound sources in movement have been conducted on intelligibility.

Weissgerber et al. [245] assessed SRM for adults with normal hearing and cochlear implant (CI) users, employing a speech-in-noise test with a virtual noise source in movement. The target speech was located at $0°$ azimuth while the masker was moving counter-clockwise from $90°$ to $270°$ azimuth (i.e., from left to right) at a constant radius of 0.7 meters. The playback system included 128 loudspeakers and the virtual masker sound source was created by including time delays and level adjustments between the loudspeakers. For subjects with normal hearing, the authors reported an SRM of 4.3 dB (standard deviation of 1.6 dB) for continuous noise (continuous broadband noise with the same long-term average spectrum as the target speech). No other movement was reported covering, for example, azimuth angles in the frontal quadrant of the frontal plane.

Davis et al. [71] evaluated the percentage of correct answers for adults with normal hearing through a speech-in-noise test with moving sound sources. The target and maskers were speech, recorded for the same person (male). They assessed seven experimental conditions: two involved target motion, two involved maskers motion, one condition involved a sudden change in target position, and two co-located condition without movement as control conditions. The motion of sound sources was simulated via amplitude panning between two adjacent loudspeakers in a 64-channel circular loudspeaker array. The authors conclude that even small momentary spatial separations at the moment of present the keywords in the moving cases, led to better performance than with stationary

cases.
Pastore and Yost [192] evaluated speech intelligibility in adults with normal hearings, comparing SRTs between conditions where the target was moving to those where it was stationary. As previous studies determined that the comprehension of the first part of the target sentence is relevant to understand the second part [7, 71], they used a one-word target to eliminate this effect, thus examine how the movement alone affects speech intelligibility. Four conditions with two or four maskers were tested: target and maskers co-located at $0°$ azimuth, two maskers at $±45°$ azimuth, and four maskers at $±45°$ and $±90°$ azimuth. Only the target was moved between $±20°$ azimuths. To simulate the target's movement, amplitude panning was applied to the loudspeakers at $±30°$ azimuth. A non-significant difference was found between the moving and stationary conditions, in terms of percent correct keywords and SRM. These results differed from the ones reported by Davis et al. [71]. Nevertheless, there are many differences between the two studies and any of those differences could lead to different results. The probable main distinguishing factor in Davis et al. [71] was the length of the target source, enabling easier understanding. No masker in movement condition was tested.

4

Experimental Setup

All the experiments performed in this thesis took place in a sound-attenuated listening booth at the Institute of Technical Acoustics (ITA) Aachen, which has a volume of V ≈ 10.5 m^3 (l x w x h $[m^3]$ = 2.3 x 2.3 x 2.0) (Figure 4.1). Further details on room acoustics inside the booth can be found on Pausch et al. [194]. Participants were seated in front of a display and they were able to use a keyboard for data input. A graphical user interface (GUI) and test routine were developed in Matlab (The MathWorks Inc., Natick, MA) to playback test stimuli, record and evaluate responses, and perform the adaptive adjustment of SNRs (see section 3.2). For the audio playback, an auralization based on a binaural reproduction technique was used (see section 2.3). The auralized signals were presented through a pair of headphones (Sennheiser HD 650, Wedemark, Germany) (Figure 4.2). With the use of headphones, the sound is directly played back over the external ear into the ear canal and the eardrum. Therefore the influence of the headphones, external ear and ear canal, was compensated in order to obtain an accurate reproduction of the sound signals at both eardrums (see subsection 2.3.3). The stimuli presented were audio-only without other help, for example, from visual cues.

A set of HRTF was convolved with speech stimuli to be rendered in free-field and reverberant conditions. To reproduce the virtual acoustic conditions, two software developed at ITA were used.

4.1 Acoustic Virtual Reproduction Software

For the free-field conditions, the virtual acoustic scenes were created using the software Virtual Acoustics (VA) [114]. Positions of the listener and the target sound source, as well as the moving trajectory of the maskers, were defined in VA. The target sound source was always located at 0° azimuth relative to the listener

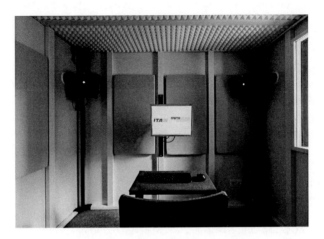

Figure 4.1: ITA sound-attenuated hearing booth

Figure 4.2: High quality headphones Sennheiser HD 650

in VA. The relative distance between the listener position and the positions of the target and maskers sound sources was fixed at 1 m to avoid biases due to the Doppler effect.

The reverberant simulations were based on the software library RAVEN [217, 218]. The interfaces of the RAVEN module include functions to define the scene and run simulations in various configurations. Results of the RAVEN simulation models have been validated for various conditions [10, 196]. RAVEN defined state-of-the-art algorithms and it includes hybrid acoustic simulation models to generate single components of a room impulse response (RIR). This simulator is also able to implement directivity for the sound sources using DAFF open source format. Directivity data can be loaded from OpenDAFF database files, which support octave and third-octave frequency resolution and arbitrary spatial resolution [244].

For the reverberant conditions a shoebox room with the dimensions 7 m x 6 m x 2.8 m (V= 117.6 m^3; S= 156.8 m^2) was simulated. To all six surfaces of the room, absorption and scattering coefficients were applied homogeneously. The scattering coefficients had a frequency independent value of 0.1. The absorption coefficients were frequency dependent, leading to different values for the reverberation time (T20, averaged for the 500 Hz and 1 kHz frequency band) between 0.5 s and 1.31 s. The evaluation of the reverberation times was done according to ISO 3382 − 2 [117] using the *ita_roomacoustics* function [43] of the ITA-Toolbox [18], an open-source project for Matlab. The target sound source was always located at 0° azimuth relative to the listener, at distance of 1.5 m. To account for the directionality of the human voice, a source directivity dataset was applied to the sound sources [133]. For the receiver, a HRTF dataset of an artificial head developed at ITA [214], with a resolution of 3° in both azimuth and elevation angles, was assigned. Each HRTF had a length of 256 samples, using a sampling rate of 44,100 Hz.

4.2 Dynamic Binaural Reproduction

In the test conditions when listeners' head movement was allowed, the binaural auralization was conducted in real-time during the experiment by updating the HRTFs based on the listener position in the virtual scene. An array of four optical tracking cameras (Flex13 models from OptiTrack, NaturalPoint, Inc., Corvallis, OR, USA, see Figure 4.3) inside the listening booth was utilized to track a rigid body attached on top of the headphones worn by the participant to capture the head movement at a sampling frequency of 120 frames per second and to update

Figure 4.3: Camera OptiTrack Flex 13, with 1.3 million pixels of resolution and 120 FPS sample rate

the listener's position in VA (Figure 4.4).

The same set of HRTFs measured from the artificial head developed at ITA [214], with a resolution of 3° in both azimuth and elevation, was used in the binaural auralization to render the sound sources in a free-field condition. Instead of interpolating between HRTFs for head angles located between the measured HRTF angles that increased overall system latency, the listener head position was rounded to the nearest angle with which the HRTFs were updated.

Participants were seated in the middle of the room that was completely dark, both windows closed, with the purpose that listeners could focus only on the acoustic stimuli. To prevent that the attention of listeners focused on the screen, thus restricting their head movements, no display in front of listeners was used and all answers were given out loud. To ensure that participants were not conscious of changes in the head tracking condition, they were asked to wear the head tracker cap during the whole test session.

43

Figure 4.4: Headphones Sennheiser HD 650 with the rigid body on the top

5

Digit Triplet Test

One of the main purposes of this thesis was to assess speech intelligibility by means of a listening test that can represent more realistic everyday situations. Also important is to consider possible clinical uses, for that reason, the speech material must represent, as close as possible, a daily conversation but must be short enough so it can be plausible its use as a clinical test. For a more realistic scenario, the use of sentences or continuous speech could be more appropriate, but the time constraints that a normal clinical assessment has, lead to choosing short stimuli. To achieve the premises of this thesis, it was decided to use the digit-triplet speech-in-noise test [222]. The reason is that digits are among the most frequent words and therefore very familiar (real situations), moreover, triplets would give more accurate results than using single words. Additionally, the use of digits made it possible to make the test completely automatic since the responses can be given by pressing the digit keys on any keyboard.

Currently, the digit-triplet test in the German language was already developed by Zokoll et al. [265]. They created six different test lists of 27 triplets, thus, only six different cases could be assessed without repeat the lists due to the possible learning effect. Nevertheless, in this thesis more than 6 cases were evaluated during each experiment. Therefore, a new set of German digit-triplets was developed with the purpose of increasing the number of different test lists. Following, all the different steps followed for the construction of the test are presented.

5.1 Construction

The new set of German digit-triplet test was developed based on methods from Smits et al., Zokoll et al. and the International Collegium of Rehabilitative Audiology (ICRA, working group on multilingual speech tests) [3, 223, 265]. One trial of the test is composed of three digits between zero and nine in one utterance.

The test comprises several lists to assess different cases or scenarios and each list contains each digit three times at the respective position in the triplet.

Although choosing digits as speech material seems simple, several parameters of the test have to be selected that are specific for each language, e. g. the number of syllables of the digits, type of speaker (gender, training, pronunciation), recording procedures, and type of masker used.

5.1.1 Word selection

To build a digit-triplet test it is possible to choose between ten digits (0 to 9) but, depending on the language, not all of them can be used. To avoid certain digit being recognized purely by its unique number of syllables, the number of syllables of each digit needs to be considered to maintain the homogeneity of the speech material. For example the Dutch digit-triplet test [222] was created using only the eight monosyllabic digits, for the British English digit-triplet test [100], the nine monosyllabic digits were selected, but for the Polish digit-triplet test [187], the ten digits were selected since the proportion of monosyllabic and disyllabic digits were similar (4 and 6 respectively).

This thesis was developed in the German language and only one of the ten digits contains more than one syllable (7, "sieben" in German), for that reason, were choose the nine monosyllabic digits [265].

Previous to the presentation of the speech material, a short pure-tone signal of 1 kHz was playback to focus the listeners' attention to the following three digits (keywords of the test).

5.1.2 Speaker

The digits were recorded by a 21 years old female German native speaker using a normal intonation and constant vocal effort. The speaker's fundamental frequency was 237.6 Hz and the recordings took place in the semi-anechoic chamber of RWTH Aachen University[1]. The speaker's mouth and the recording microphone were separated by approximately 1 m of distance to avoid significant reflections. To facilitate the test resynthesis, short but distinctive pauses between successive digits were made.

---

[1] The digit-triplet corpus developed in this thesis is open source. DOI: 10.1109/RWTH-2019-06230.

5.1.3 Recording

For the recording of the triplets, three lists of nine different triplets were composed containing only once each digit at each position. In this way, the digit-triplets were created by recording each digit in all three possible sequential positions and artificially concatenating them with a 200 ms inter-stimulus interval. Thus, the digit-triplets were created having normal intonation and natural prosody for the resynthesized material. All recordings were done at a sampling rate of 44.1 kHz and a resolution of 16 bit.

5.1.4 Resynthesis

With the purpose of creating different lists, each digit at each position in the recorded triplet was cut, omitting the pauses, and resynthesized into new triplets in a way that preserves the prosody. For each triplet was used the first-position recordings for all first, second-position for all second, and third-position for all third. A pause between successive was added. This procedure allows optimization of the speech material by adjusting the SRT of each digit in each position to match the mean digit-wise SRT as closely as possible [265].

5.1.5 Masking noise

The masker was a randomized superposition of all digit-triplets used in the list, with a random delay up to 4 s between successive repetitions of the speech items. This resulted in a quasi-stationary noise with the same long-term averaged spectrum as the target speech. The spectral distribution of the masker is shown in Figure 5.1.

A masking noise having the same long-term spectrum as the speech material is recommended [179, 223], since the highest efficiency in energetic masking is obtained by a spectral match between the average target speech and the masker [108]. For that reason, with the use of another type of masker such as white noise or any other speech-shaped noise built from a different speech material, it would not be obtained the highest efficiency in energetic masking.

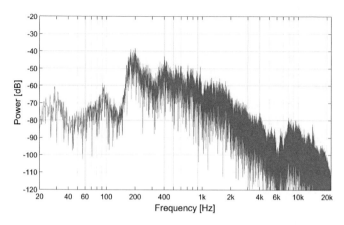

Figure 5.1: Noise power distribution per frequencies between 20 Hz and 20 kHz.

5.2 Optimization

Measures had to be taken to optimize digit triplets and to ensure equal intelligibility across the speech material. The optimization was determined by a speech intelligibility measurement of the resynthesized speech material at fixed SNRs, covering the range of 10 % to 90 % in speech intelligibility at a masker level of 65 dB SPL.

A total of 12 young adults (5 female, 7 male) participated in the optimization test, aged between 21 and 28 years. All participants had normal hearing (pure tone thresholds < 20 dB hearing level between 125 and 8000 Hz) at the time of the experiment and they speak German as their native language. Participants were seated in the ITA sound-attenuating booth in front of a screen and used a keyboard for data input (see chapter 4). The speech material of the test consisted of 27 triplets, where each digit was either in the first, second, or third position in the triplet. The 27 triplets were tested at seven different SNRs: -2, -4, -7, -10, -13, -16 and -18 dB. The listeners were asked to recall the three digits of each trial using the keyboard. In the case when the listener cannot recall one or more digits, they were allowed to leave the space blank.

The purpose of the optimization measurement was to obtain a digit-wise intelligibility function for each digit played back at each position. For that reason digit scoring was used, that means the score as correct or incorrect was taken into account for each digit in each position of the triplet. A level adjustment for each digit in each position was made to reach the SRT at 50% of intelligibility. If the SRT of a certain digit is not close to the mean digit-wise SRT, an adjustment up to 4 dB could be made. If the necessary adjustment is bigger than 4 dB, that digit in that specific position of the triplet could be eliminated. In this thesis two cases were reported out of these boundaries: digit eight in the first position and digit four in the third position.

5.3 Evaluation

After the optimization procedure, 14 lists containing 24 digit-triplets were created. In all lists, each digit in each position appears three times, but all triplets in all lists are unique, which means that none triplet was repeated.

A second listening test was performed to derive the psychometric function of the new digit-triplet test that will be used as speech-in-noise materials. A subset of seven lists was randomly chosen to be tested under three SNRs corresponding about 20 %, 50 % and 80 % of speech intelligibility at a masker level of 65 dB SPL.

A total of 18 young adults (9 female, 9 male) participated in the evaluation test, aged between 18 and 36 years. All participants had normal hearing (pure tone thresholds < 20 dB hearing level between 125 and 8000 Hz) at the time of the experiment and they speak German as their native language. The listeners who participated during this phase were different from those who participated in the previous optimization test. The speech material involved seven lists of 24 triplets and it was tested at three different SNRs: -7, -9 and -11 dB. The listeners were asked to recall the three digits of each trial using the keyboard. In the case when the listener cannot recall one or more digits, they were allowed to leave the space blank.

The purpose of the evaluation measurement was to obtain a triplet-wise intelligibility function for each digit-triplet. For that reason triplet scoring was used, that means, the three digits of the triplet must be correctly answered to count as a correct response. A logistic model function (Eq. 5.1) was fitted to the speech intelligibility data of all participants, with the purpose of obtaining the triplet-wise psychometric function [265].

$$Speech\, Intelligibility\,(SNR) = y + (1 - y) * \frac{1}{1 + e^{(4s(SRT - SNR))}} \quad (5.1)$$

In the Eq. 5.1, SNR correspond to the signal-to-noise ratio, y is the chance level (in this case 0.1, since there are ten alternatives for all digits on the keyboard)

[223, 265], SRT is the mean speech reception threshold (SNR at 50 % of speech intelligibility) and s represent the slope at the SRT.

The digit-wise and triplet-wise psychometric functions of the subset of seven unique lists are reported in Table 5.1, alongside that from Zokoll et al. [265]. The mean SRT reported was calculated from SRTs interpolated from the psychometric functions at 50% intelligibility performance. With a similar mean slope but lower mean SRT, the psychometric functions of the seven new digit-triplet lists share similar shapes but overall better intelligibility by 2.6 dB than those from Zokoll et al. The improvement on the intelligibility of the new digit-triplet test could have been due to the use of a different voice for the recordings, the use of a normal intonation and natural prosody, or individual differences of listeners. The subset of the new digit-triplet test was deemed validated since the mean slope and standard deviation remain similar to that of Zokoll et al. Although no psychometric function was explicitly derived with listeners due to prolonged experiment duration, the other half of the digit-triplet lists shall share similar properties as the seven lists shown in Table 5.1 under the assumption that all lists were created and optimized using the same procedure.

Table 5.1: Parameters of psychometric functions in the digit-triplet test created by [265] and from the present thesis, both digit- and triplet-wise scoring. The mean SRT of 50% intelligibility and the mean slope of the psychometric functions are listed with standard deviation σ.

	SRT $\pm$ σ [dB]	Slope $\pm$ σ [%/dB]
Zokoll et al. (2012), digit-wise	-11 $\pm$ 0.2	14.5 $\pm$ 1.3
Zokoll et al. (2012), triplet-wise	-9.3 $\pm$ 0.2	19.6 $\pm$ 2.2
Present thesis, digit-wise	-12.2 $\pm$ 0.3	12.8 $\pm$ 1.8
Present thesis, triplet-wise	-11.9 $\pm$ 0.2	18.8 $\pm$ 2.1

5.4 Sequence presentation

The presentation sequence of a typical trial is illustrated in Figure 5.2. In each trial, the duration of the masker was constant at 4 s. The masker stream started 500 ms before the target stream onset. The target stream, always located at 0° azimuth, began with a 750 ms leading pure tone (1 kHz). The leading pure tone was always presented at 3 dB louder than the target stimuli. The inter-digit interval was variable within each digit-triplet. Since the shortest and the longest digit-triplet was 1750 ms and 2100 ms, respectively, variables silent times between 350 and 250 ms were used. Thus, when the masker was circularly moving, the midpoint of the first digit was played back at 1.8 s ± 0.2 s, the midpoint of the second digit was played back at 2.6 s ± 0.2 s, and the midpoint of the third digit was played back at 3.4 s ± 0.2 s.

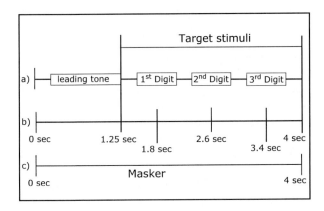

Figure 5.2: Sequence of the digit-triplet test. (a) Illustrations of the digit-triplet stimulus playback stream. (b) Mean playback time of all digits in each position of the triplet. (c) Timeline of the noise masker.

6

Dynamic Speech-in-Noise Test

In real-life environments, people often listen to speech mixed with distracting masking noises. The maskers could be stationary such as a blender machine or a microwave, but generally, the maskers are moving around us. Only a few speech perception studies have evaluated moving sound sources (see section 3.6). Most current research has focused on assessing speech-in-noise perception for stationary sound sources (see chapter 3), despite real-life environments are constantly moving. Therefore, due to the variety of moving maskers in everyday situations, evaluate SRM (see section 3.3) of a moving masker was considered of utmost importance. In this experiment, the SRM calculation for moving maskers is to be understood in terms of dynamic spatial release from masking (DSRM).

The movements of the maskers are unpredictable. It is possible to find different types of trajectories such as radial movements (a car driving away in a straight line), circular movements (a waiter walking around our table in a restaurant) or a completely random trajectory. In order to facilitate the analysis, only circular movements at a constant radius were assessed in this section. Keeping a constant radius throughout the entire movement of the masker, the sound pressure level (SPL) at the center of the circumference (position of the listener) remains constant, thus facilitating the analysis. Despite the simplification of the masker's movement, the masker's trajectory could be clockwise or counterclockwise. Therefore two different masker trajectories, with the target always at the front of the listener in $0°$ azimuth, were evaluated:

Masker trajectory away

The movement of the masker starts at the same position as the target (co-located) and then moves away from the target position to a different azimuth angle.

Masker trajectory toward

The other possibility is that the masker starts its movement in any azimuth angle (e.g. in $90°$) and then moves toward the target position at $0°$.

Evaluating several angular movements, both away and toward the target, it will be possible to bring insight into spatial separation benefits of a moving masker. Allen et al. [7] carried out a similar experiment. The stimuli were taken from the coordinated response measure (CRM) corpus [26]. It consist of sentences which take the form "Ready [call sign] go to [color] [number] now", where the call sign consisted of the target identifying "Baron" or one of the masking call signs ("Charlie", "Ringo", "Eagle", "Arrow", "Hopper", "Tiger" and "Laker"). Colors were chosen from red, white, blue, and green and the numbers from 1 to 8. Thus, the task was identifying in the first segment the call sign and repeats the two keywords: the color and the number. The target talker was identifying by the call sign "Baron" and the participants have to identify both of the keywords following the call sign in the presence of two masker talkers with different call sign and scoring words. They analyzed four masker conditions:

Co-located
Target and both maskers were played back from the central loudspeaker.

Separated
Target was played back from the central loudspeaker, one masker was played back from a loudspeaker 30° to the right and the other 30° to the left from the central loudspeaker.

Start separated
Target and maskers start as in separated condition but maskers change the position to the central loudspeaker after 700 ms (just after the identifying call sign).

Start co-located
Target and maskers start from the central loudspeaker as in co-located condition but maskers change the position as in separated condition after 700 ms.

They tested two conditions with maskers stationary and two conditions with maskers changing suddenly the positions, not moving. For all participants, the SRMs (in dB) were higher for the case start separated than the co-located. This indicates that spatial separation between target and masker, even during the beginning of the stimulus only, provides a significant advantage than when both sources are co-located during the entire stimulus.

In the current experiment, the movement "toward" begins its movement separated from the target, thus, it could have an advantage with respect to the movement "away" that begins its movement co-located with the target. Therefore, the hypothesis is that the masker moving toward will show greater SRM than the masker moving away.

The aims of this experiment were (1) to investigate the SRM benefits of a moving masker (DSRM) and (2) compare between two trajectories (away and toward) if listeners reach different DSRM benefits.

6.1 Experimental Methodology

A total of 28 young adults (13 female), aged between 19 and 36 years, completed the listening experiment. All participants had normal hearing (pure tone thresholds < 20 dB hearing level between 125 and 8000 Hz) at the time of the experiment and they speak German as their native language. Each participant was provided instructions regarding the tasks and they gave written consent prior to testing.

Speech stimuli samples were a set of German digit-triplet tests (see chapter 5). The masking noise was a randomized superposition of all digits used in the test (see subsection 5.1.5), resulted in a quasi-stationary noise with the same long-term averaged spectrum as the target speech. For the audio playback, an auralization technique based on a binaural reproduction was used (see section 2.3).

The test was carried out in the ITA sound-attenuated listening booth (see chapter 4). The target stream of digit-triplet is defined in section 5.4. The listeners were asked to recall the three digits of each trial using the keyboard. In the case when the listener cannot recall one or more digits, they were allowed to leave the space blank.

Testing with each subject was organized into 11 cases for which 11 lists of digit-triplets were used. Each participant performed one case with masker stationary at $0°$ (baseline for all DSRM calculation) and ten cases with masker moving away or toward the target position in: $15°$, $30°$, $45°$, $60°$ and $90°$. Figure 6.1 provides a graphical illustration of all the cases.

Each digit in the triplet was scored as correct only when the digit itself and the sequential position were correctly identified. Possible scores of each digit-triplet trial were 0 %, 33 %, 66 %, and 100 %. A simple up-down adaptive procedure (see section 3.2) was used to track the SNR at 50 % speech intelligibility by

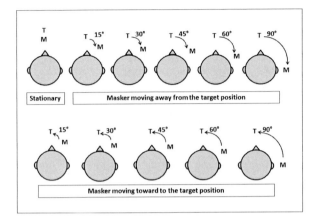

Figure 6.1: Graphical representation of all cases in this study and the masker
position/movement. Five different masker movements (15°, 30°,
45°, 60° and 90°) for two trajectories (away and toward the target
position) were evaluated. T denotes the target sound source location
and M is masker position/movement during the trial.

changing the target level while keeping the masker level at 70 dB (re 20 μPa).
The initial masker level was played back at 70 dB (re 20 μPa), resulting in 0 dB
SNR. The initial step size was set to 4 dB until the first reversal was reached,
from which on the step size was 2 dB. Each test case finished when participants
reached sixth reversals or the twenty-fourth trial that is the maximum number
of trial in each list[2]. The SRT was calculated of the average from the last four
reversals.

6.2 Results

A repeated-measures analysis of variance (ANOVA) was fitted to the SRT data
with masker conditions (stationary at 0° vs. moving away 15° vs. moving
toward 15° vs. moving away 30° vs. moving toward 30° vs. moving away
45° vs. moving toward 45° vs. moving away 60° vs. moving toward 60° vs.
moving away 90° vs. moving toward 90°) as the within-subjects variable. The
ordinate of Figure 6.2 shows mean SRT data in decibels over different masker

---

[2]None of the participants reached the twenty-fourth trial.

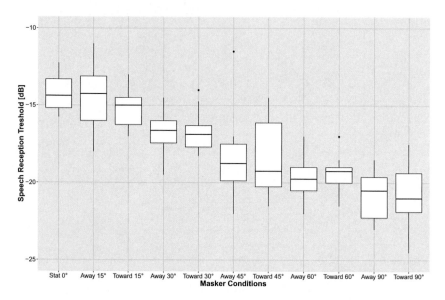

Figure 6.2: Speech reception threshold (SRT) measured for eleven different masker conditions: stationary at 0°, and moving away and toward the target position (15°, 30°, 45°, 60° and 90°).

conditions on the abscissa. The ANOVA revealed a significant main effect of masker conditions [$F(10,130) = 35.55$, $p < .001$], showing higher SRTs for shorter degrees of separation than for larger degrees of separation between the target and masker. A Bonferroni-corrected pairwise comparison was applied to examine possible differences between the movement of the masker away or toward the target position, nevertheless the differences were no significant ($p > .05$) for all five degrees of movements (see Figure 6.2). Figure 6.3 shows the measured DSRM for masker trajectories away and toward the target position across various movements.

DSRMs were calculated by subtracting the SRT from masker at stationary 0° from the SRT in each masker trajectory. The DSRM data has the following factors: movement of the masker (five levels: 15, 30, 45, 60 and 90 degrees) and masker trajectory (two levels: away and toward). A significant main effect of movement of the masker [$F(4,52) = 51.38$; $p < .001$] was found, but a not significant main effect in the masker trajectory [$F(1,13) = 0.04$; $p > .05$]. The interaction between the two factors also was not significant [$F(4,52) = 0.31$; $p > .05$]. As seen in Figure 6.3 movement away or toward the target, resulted in similar DSRM across all five movements.

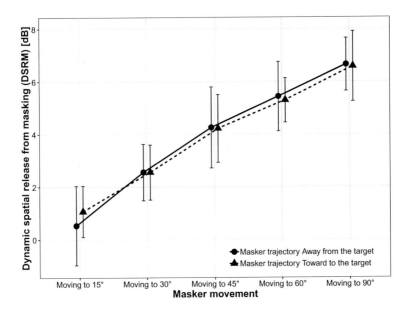

Figure 6.3: Dynamic spatial release from masking (DSRM) measured for five different masker movements (15°, 30°, 45°, 60° and 90°) and for two trajectories: away and toward to the target position.

A pairwise comparison with Bonferroni correction was used to examine more closely the main effect of movement. Only the relation between 45° and 60° was found not significant different ($p > .05$). All the remaining relations were found significant different ($p < .001$) with masker moving away from the target (1) 15°, mean (M) = 0.53, standard deviation (SD) = 2.4, (2) 30°, M = 2.55, SD = 1.7, (3) 45°, M = 4.25, SD = 2.4, (4) 60°, M = 5.42, SD = 2.1, and (5) 90°, M = 6.64, SD = 1.6. Due to the similar results between away and toward movements, only the mean results of the masker trajectory away were reported.

6.3 Discussion and Conclusions

An experiment to assess DSRM with a masker moving in several azimuth angles was done. Two masker trajectories and five masker movements were evaluated only in the frontal quadrant of the frontal plane, with a maximum movement of 90°.

The DSRMs results showed a progressive increment of benefits (in dB) as the masker's movement became longer. Nevertheless, studies of SRM with stationary maskers reported greater benefits than those measured in this experiment. For example, the measured masker movement of 90° has a mean DSRM of 6.64 dB and for a stationary maskers at 90° a SRM of around 9 dB was reported [34, 121]. This difference was expected, due to the spatial separation benefits for a masker in movement is variable, unlike a stationary masker that have constant benefits during the entire trial. In the next chapters, several comparative analyses between moving and stationary maskers, under different auditory conditions, are shown. The comparative analysis has the purpose of evaluating if the different SRM between moving and stationary maskers are caused solely by the movement. No significant difference was found between trajectories away and toward the target position. This finding is not consistent with those from Allen et al. [7]. However, there are many differences between the two experiments. One important difference is the type of the masker. Meanwhile, Allen et al. used speech as a masker, in this experiemnt a quasi-stationary noise with the same long-term averaged spectrum as the target speech was used. Allen et al. suggested that initial spatial separation (comparing start separated versus co-located cases) could act as a prime aiding identification of maskers and target using nonspatial differences in the talkers. This advantage could have been unavailable in the current experiment since the masker was not speech.

7

Dynamic SRM: Binaural and Monaural Contributions

It is well known that listeners have a greater capability to identify two (or more) sound sources as different if the sources are spatially separated than co-located [65, 105, 134]. This ability is helpful in many social situations where listeners are exposed to several sound sources at the same time such as a crowded street or a full restaurant. When speech is presented together with a noise masker, it is possible to gain an advantage in speech intelligibility if the masker is located at a different position than co-located with the target (see section 3.3).

Many factors contribute to improve or reduce SRM including the measurement paradigm [34], room acoustic conditions [68, 124, 147, 247], type of masker [40, 85], the number of maskers in the scene and its spatial distribution [39, 104]. However, most studies have focused on maskers that are located at stationary positions, therefore, the effect of a moving masker has not yet been well documented.

Recent SRM studies, assessing moving versus stationary maskers, demonstrated that even a small momentary spatial separation led to better performance in speech-in-noise perception (see section 3.6). For that reason, even with so few SRM studies with moving sound sources, it is possible to find support to expect differences in SRM between moving and stationary sound sources. Consequently, the movement could be another factor to consider in a SRM evaluation.

Currently, there are several models to predict SRM. The models are suitable for different types of maskers and for multiple masker locations in the horizontal plane, however, they may not be suitable for predicting DSRM. Both mathematical models described in section 3.4 have been successfully evaluated predicting SRM for several maskers in stationary positions [67, 169], for that reason, examine their validity in cases with moving maskers is worthwhile. In the following text, both models [34, 121] are to be understood in terms of the *stationary models*.

To examine whether the *stationary models* can predict DSRM, a comparison between predicted SRM and DSRM must be carried out. It was considered that drawing a direct comparison between stationary and moving masker conditions is not accurate (e.g., SRM of a stationary masker at 90° azimuth versus SRM of a moving masker from 0° to 90° azimuth) because, for a stationary masker, the SRM is constant during the entire stimulus but for a moving masker the SRM is continuously changing in time. For that reason, the SRM predicted by the *stationary models* were calculated taken the accumulated SRM from 0° (target position) to a certain θ degree of separation (see equation 7.1). Therefore, the DSRM was compared with the accumulated SRM predicted by the *stationary models*.

$$SRM_{accumulated} = \int_{0}^{\theta} SRM(\theta)d\theta \qquad (7.1)$$

Taking into account what was mentioned in subsection 2.2.5, the SRM is compose of the sum of two components: monaural advantage that arises from the improvements in SNR at the better-ear, and binaural advantage that arises from the binaural unmasking (especially by ITDs).Therefore, a two-masker configuration, moving away from the target, are measured to assess the contribution of each DSRM component:

Binaural component
The binaural contribution is possible to determine by measuring DSRM in a bilaterally symmetric two-masker configuration, in which one masker is moving from 0° to θ degrees to the left of the target and the other masker is moving from 0° to θ degrees to the right of the target. Under this condition the better-ear effect is cancel and only the binaural contribution is measured [104, 121]. For the following, this bilaterally symmetric two-masker configuration will denote as the "$-\theta/+\theta$" configuration:

$$Binaural\ contribution = SRM(-\theta/+\theta),$$

θ represents the reached final degree at the end of the movement of the masker, negative values are movements to the left; and positive angles are movements to the right. In this section, all trajectories of maskers are circular movements maintaining the same radius during the entire stimulus.

In consequence, for cases with moving maskers will be:

$$Binaural\ contribution = DSRM(-\theta/+\theta) \qquad (7.2)$$

Better-ear component

The contribution of the better-ear effect could be examined by measuring DSRM in an asymmetric masker configuration of the form "$+\theta/+\theta$", in which the two maskers are moving together in the same direction. Nevertheless, with this configuration the binaural contribution is not canceled. For that reason, to examine the better-ear contribution only, the binaural contribution calculated with the masker configuration $-\theta/+\theta$ must be subtracted:

$$Better - ear\ contribution = DSRM(+\theta/+\theta) - DSRM(-\theta/+\theta) \quad (7.3)$$

The target playing back digit triplets was always positioned in front of the listener at $0°$ azimuth in the virtual scene. The two maskers played back the quasi-stationary noise described in subsection 5.1.5.

A DSRM analysis with maskers moving on different trajectories could, therefore, bring insight into dynamic binaural speech intelligibility. Therefore, the aims of this experiment were (1) to investigate the respective DSRM better-ear and binaural contributions, and (2) examine whether the *stationary models* can predict DSRM.

7.1 Basic Concepts

The auditory perception of stationary stimuli was based on psychophysical studies as evidenced by Weber, Fechner, and Stevens laws [11, 79, 163, 232]. With the inclusion of moving sound sources in the current studies, it would be necessary to review some concepts to be applicable for stimulus in movement.

To know how to assess speech perception for moving sound sources, it is necessary to review the concepts largely applied on stationary stimuli, thus, knowing if it is possible to extend them for moving cases also.

How was explained before in chapter 3, one of the most widely parameters used to assess speech perception is the SRT and the SRM. Plomp [204] defined SRT as the SNR in decibels at which the 50 % of the speech material (words or sentences) is correctly identified. Calculating the SRT requires a psychophysical analysis that relates the perceived proportion of words (speech target) to a physical ratio (SNR between target and masker) as presented by Gelfand et al. [86]. The difference for moving sound sources is that the SNR is not constant during the entire stimulus. Nevertheless, our interpretation is that the SRT represents the

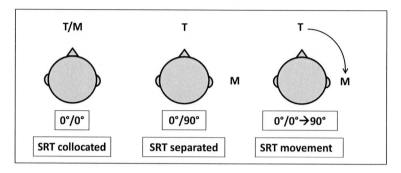

Figure 7.1: Target/Masker (T/M) configurations to calculate SRM. Masker sound sources were either located at stationary positions or moving along a trajectory.

capacity to understand speech under any masker condition, no matter if the masker is stationary or moving, at 50 % intelligibility.

Davis et al., Pastore and Yost, and Weissgerber et al. [71, 192, 245] assessed stationary and moving sound sources in a speech-in-noise task (see section 3.6). In these experiments, SRT, SRM, and percent of correct answers were evaluated as a measurement of speech-in-noise perception. Although they did not report it specifically, it is possible to assume that their interpretation was similar to that in this thesis. They used the same procedure indistinctly for stationary or moving sources to calculate SRT or percent of correct answers. This fact points out that the goal is to know how much is possible to perceive in a speech-in-noise test, no matter the condition of the masker.

The SRM is computed as: $SRM = SRT_{collocated} - SRT_{separated}$ (see Figure 7.1) for a stationary masker. Therefore, due to the previous explanation, it is also possible to extend this concept for moving maskers, defining the DSRM as:

$$DSRM = SRT_{collocated} - SRT_{movement} \qquad (7.4)$$

$SRT_{collocated}$ represents the static case where the masker is collocated with the target over the entire stimulus (typically at 0° azimuth), whereas $SRT_{movement}$ represents any case with a moving masker. Also it could be applicable for a target in movement, but in this thesis only the masker moves.

7.2 Experimental Methodology

A total of 14 young adults (4 female, 10 male) participated in the test, aged between 21 and 39 years. All participants had normal hearing (pure tone thresholds < 20 dB hearing level between 125 and 8000 Hz) at the time of the experiment and speak German as their native language. Each participant was provided instructions regarding the tasks and they gave written consent prior to testing. Speech stimuli samples were a set of German digit-triplet tests (see chapter 5). The masking noise was a randomized superposition of all digits used in the test (see subsection 5.1.5). It is important to mention that 48 independently generated maskers were used in this experiment. The computer routine randomly selects two of the 48 maskers for been playing back in the first trial. For the second trial the two maskers previously used were not available anymore and another two randomly maskers were selected. Because each list contain a maximum of 24 trials, none pair of maskers was repeated. This routine started over for each list (each case tested).

For the audio playback, an auralization technique based on a binaural reproduction was used (see section 2.3).

The target stream of digit-triplet is defined in section 5.4. Once both the target and masker streams finished, a GUI window appeared for participants to enter the digit triplet (see chapter 4). The listeners were asked to recall the three digits of each trial using the keyboard. In the case when the listener cannot recall one or more digits, they were allowed to leave the space blank.

Testing with each subject was organized into 12 cases for which 12 lists of digit-triplets were used. The order of the cases was balanced across subjects in a Latin Square design. All tested cases are shown in Figure 7.2. Two stationary cases were assessed as control measurement: two maskers at 0° (baseline for all DSRM calculation) and two maskers at 90° to the right of the listener (to compare with the *stationary models*). Ten moving two-masker configuration were assessed in this experiment. Five cases to evaluate DSRM($-\theta/+\theta$) and five cases to assess DSRM($+\theta/+\theta$) were used to examined the binaural and better-ear contributions (see equations 7.2 and 7.3). The angular movements of maskers always started from 0°.

Raw DSRM results were used to calculate:

DSRM($+\theta/+\theta$): The masker array was asymmetrical, with both maskers moving together to 15°, 30°, 45°, 60° or 90°.

DSRM($-\theta/+\theta$): The masker array was symmetrical, with one masker moving to the left and the other to the right at the same time, to 15°, 30°, 45°, 60° or 90°. With this configuration was obtained the binaural contribution (see equation 7.2).

DSRM($+\theta/+\theta$)-DSRM($-\theta/+\theta$): With this subtraction is obtained the better-ear contribution (see equation 7.3).

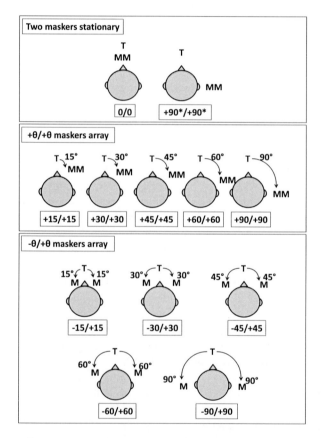

Figure 7.2: Graphical representation of all conditions tested in the current experiment showing the masker's location/trajectory. T represents the target sound source and M is the masker source position or movement during the trial. * represent both maskers stationary at +90°

Each digit in the triplet was scored as correct only when the digit itself and the sequential position were correctly identified. Possible scores of each digit-triplet trial were 0 %, 33 %, 66 %, and 100 %. A simple up-down adaptive procedure [154] was used to track the SNR at 50 % speech intelligibility by changing the target level. The initial masker noise was played back at 70 dB (re 20 μPa), resulting in 0 dB SNR. The initial step size was set to 4 dB until the first reversal was reached, from which on the step size was 2 dB. Each case finished when participants reached sixth reversals or the maximum trial number of twenty-four (none of the participants reached the maximum trial number). The SRT was calculated using the MATLAB psignifit toolbox (version 3.0), applying the methods described by Wichmann and Hill [249, 250].

To examine whether the *stationary models* can predict DSRM, a comparative analysis was made between current DSRM results and the accumulated SRM prediction of the *stationary models*. Figure 7.3 shows the SRM prediction of both *stationary models* when target is at 0° and the masker change the location from 0° to 90°. At the same time, the figure shows the accumulated SRM prediction of both *stationary models* when the masker move from 0° to 90°.

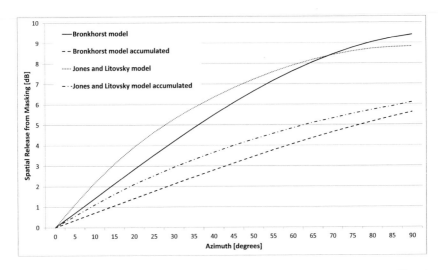

Figure 7.3: SRM prediction for both *stationary models* when target is at 0° and the masker at different locations between 0° and 90° together with the accumulated SRM of both *stationary models* when the masker move from 0° to 90°.

7.3 Results

According to Bronkhorst [34], and Jones and Litovsky [121], the predicted SRM for a stationary masker separated by $90°$ from the target source is 9.4 dB and 8.8 dB (for a noise masker), respectively. Results using one-sample t-tests suggest no significant deviation of the measured SRM with two stationary maskers at $90°$ ($M = 9.11$ dB, $SD = 1.6$ dB) from the predicted values of Jones and Litovsky's model, $p > .05$, and Bronkhorst's model, $p > .05$. Hence, it is reasonable to conclude that SRM measured in virtual acoustic environments, as done in the current thesis, aligned closely with the existing models of SRM for stationary maskers.

To analyze the contribution of both DSRM components together, a two-way ANOVA was fitted to the DSRM data in the $+\theta/+\theta$ configuration with factors: masker conditions (five levels: $15°$, $30°$, $45°$, $60°$ and $90°$) and model's prediction (three levels: prediction of Jones and Litovsky model, prediction of Bronkhorst model, and moving masker results). A significant main effect was found for both factors (masker conditions $[F(4,52) = 340.01; p < .001]$ and prediction $[F(2,26) = 6.55; p = .004]$) as well as the interactions between the two factors (masker conditions and prediction $[F(8,104) = 5.30; p < .001]$).

A pairwise comparison with Bonferroni correction was used to examine more closely the main effect of model's prediction. For the masking moving $15°$ condition, a significant difference between the moving masker results with the prediction of Jones and Litovsky model was found ($p = .03$). For the masking moving $30°$ condition, the difference was no significant between the moving masker results and both *stationary models* ($p > .05$). For the masking moving $45°$ condition, a significant difference between the moving masker results with the prediction of Bronkhorst model was found ($p = .008$). For the masking moving $60°$ condition, the difference was no significant between the moving masker results and both *stationary models* ($p > .05$). For the masking moving $90°$ condition, a significant difference between the moving masker results with the prediction of Bronkhorst model was found ($p = .01$). Figure 7.4 shows the measured DSRM for the $+\theta/+\theta$ configuration across several masker movements together with the accumulated SRM prediction for Jones and Litovsky, and Bronkhors models. The error bars show 95 % confidence intervals.

To analyze the binaural contribution only, a two-way ANOVA was fitted to the DSRM data in the $-\theta/+\theta$ configuration with factors: masker conditions (five levels: $15°$, $30°$, $45°$, $60°$ and $90°$) and model's prediction (three levels: prediction

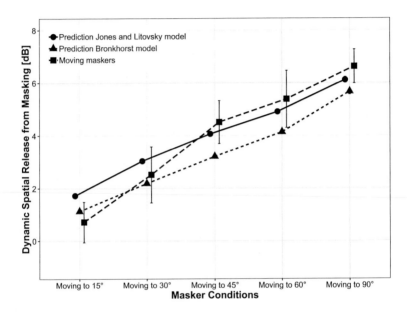

Figure 7.4: DSRM is shown for moving maskers to 15, 30, 45, 60 and 90 degrees, together with predicted SRM of both *stationary models* for the $+\theta/+\theta$ masker configuration.

of Jones and Litovsky model, prediction of Bronkhorst model, and moving masker results). A significant main effect was found for both factors (masker conditions $[F(4,52) = 20.29; p < .001]$ and prediction $[F(2,26) = 34.66; p < .001])$ as well as the interactions between the two factors (masker conditions and prediction $[F(8,104) = 3.22; p = .003])$.

A pairwise comparison with Bonferroni correction was used to examine more closely the main effect of model's prediction. For the masking moving 15° condition, the difference was no significant between the moving masker results and both *stationary models* ($p > .05$). For the masking moving 30° condition, a significant difference between the moving masker results with the prediction of Jones and Litovsky model was found ($p = .004$). For the masking moving 45° condition, a significant difference between the moving masker results and both *stationary models* was found ($p < .05$). For the masking moving 60° condition, the difference was no significant between the moving masker results and both *stationary models* ($p > .05$). For the masking moving 90° condition, a significant difference between the moving masker results with the prediction of Bronkhorst model was found ($p = .001$). Figure 7.5 serves two different functions, in addition to reporting DSRM in the $-\theta/+\theta$ configuration together with the accumulated SRM prediction for Jones and Litovsky, and Bronkhors models, also gives estimates of binaural contribution for different movements of the masker (see Eq. 7.2). The error bars show 95 % confidence intervals. To analyze the better-ear contribution only, a two-way ANOVA was fitted to the DSRM data in the configuration of equation 7.3 (subtraction between results of Figure 7.4 and Figure 7.5), with factors: masker conditions (five levels: 15°, 30°, 45°, 60° and 90°) and model's prediction (three levels: prediction of Jones and Litovsky model, prediction of Bronkhorst model, and moving masker results). A significant main effect was found for both factors (masker conditions $[F(4,52) = 153.47; p < .001]$ and prediction $[F(2,26) = 7.13; p = .003])$ as well as the interactions between the two factors (masker conditions and prediction $[F(8,104) = 2.82; p = .007])$. A pairwise comparison with Bonferroni correction was used to examine more closely the main effect of model's prediction. For the masking moving 15°, 30°, 60°, and 90° conditions, the difference was no significant between the moving masker results and both *stationary models* ($p > .05$). For the masking moving 45° condition, a significant difference between the moving masker results with the prediction of Jones and Litovsky model was found ($p = .009$). Figure 7.6 shows the measured DSRM for the masker configuration described in equation 7.3 across various masker movements, also, the accumulated SRM calculation for Jones and Litovsky, and Bronkhors models. The error bars show 95 % confidence intervals.

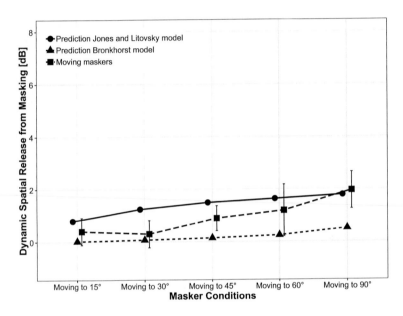

Figure 7.5: DSRM binaural contribution for maskers moving 15, 30, 45, 60 and 90 degrees, together with predicted SRM of both *stationary models*. The Binaural component is examined with the symmetric masker configuration $-\theta/+\theta$.

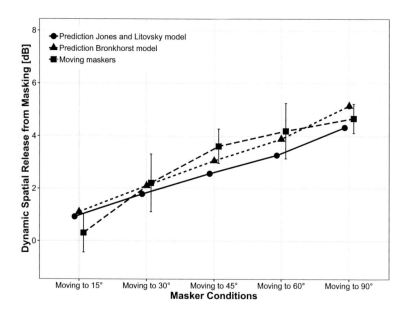

Figure 7.6: DSRM better-ear contribution for maskers moving 15, 30, 45, 60 and 90 degrees, together with predicted SRM of both *stationary models*. The better-ear component is examined with the masker configuration described in equation 7.3.

7.4 Discussion and Conclusions

The main motivations for this experiment were to assess the DSRM better-ear and binaural contributions, and examine whether the *stationary models* can predict DSRM.
In general, the comparison between DSRM and the predicted SRM of *stationary models* reveals some differences.

Binaural component: Comparing with Bronkhorst model, movements of 15 and 30 degrees were well predicted, but for movements of 45 and 90 degrees the model underpredicts the results. For the other hand, comparing with the Jones and Litovsky model, movements of 15, 30 and 90 degrees were well predicted, but for movements of 30 and 45 degrees the model overpredicts the results.

Better-ear component: Comparing with Bronkhorst model, all movements were significantly well-predicted. For the other hand, comparing with the Jones and Litovsky model, movements of 15, 30, 60 and 90 degrees were well predicted, but for movements of 45 degrees the model underpredicts the results.

The comparison between stationary and moving maskers showed that not all moving results were predicted by the *stationary models*, therefore, the movement could have a certain influence on the listener's intelligibility. A possible explanation could be the "sluggishness" of the binaural system [95]. This effect has often been modeled by a device which acts like a filter, smoothing fast changes in interaural configurations [2]. It will be necessary more research to isolate the contribution of the movement to the human intelligibility.

For future studies, it could be important to know how both binaural and monaural contributions change for larger masker movements, reaching angles behind the head. It would also be interesting to know if the DSRM have a different gain in dB when another type of masker is applied and/or a different speech target is used. Finally, a unique predictive model unifying static and moving cases could be relevant to understand dynamic binaural processes.

Listeners Head Movements in a Dynamic Speech-in-Noise Test

The head movements in speech perception tasks are been studied since Kock (1950) [134]. He found that the intelligibility increase when turning the head away from the speech source. Additional, he was the first to map out thresholds of speech-in-noise perception as a function of head orientation away from the speech target [93]. However, despite Kock's findings, several studies have proposed that for clinical trials the listener should be placed in front of the speech source, arguing that this is a more natural listening attitude [36, 135, 205].

A theoretical model proposed by Lambert [144] describes the mathematical mechanism in determining sound source location with head movement. It was later shown that non-human mammals such as cats [234] and monkeys [207] achieved better sound localization when their heads are unrestrained during the experiment. In the case of human sound localization, several studies have shown that head movements can resolve auditory ambiguities such as front-back confusion, using real and virtual sound sources [14, 118, 123, 184, 253].

While head movements suggest better localization of both target and masker sources, this could lead to improved speech perception in a speech-in-noise task, however, evidence from recent studies did not provide sufficient support for the use of such a strategy [123]. After more than 60 years since Kock's findings, Brimijoin et al. [33] were one of the first to measure head orientation during a speech-in-noise task. They evaluated the use of head orientation as a listening strategy for speech perception with asymmetric hearing-impaired participants. The stimuli were presented from one of the loudspeakers arranged in a ring set; a speech-shaped noise was used as a masker and an adaptive procedure was used to measure the SRT. They reported high variability in listeners' head movements and, in most cases, different orientation from the ideal.

Another similar study developed by Grange and Culling [93] presented a comparative analysis between a predictive model of head orientation benefits [119] and the spontaneous head orientation from listeners. The aim was to investigate if normal hearing listeners adopt an appropriate head orientation spontaneously to improve their speech perception. After the experiment is completed, a subset of participants was tested post-instructions informing about the possible benefits of head orientation. As a result, only the 56 % of the listeners spontaneously moved their head more than $10°$ (a reference from short or long head movements) and, in general, the participants did not make optimal use of their head orientation to improve their intelligibility. The predictive model in the asymmetrical cases revealed that the best head orientation is almost in between the two sound sources and the worse orientation is when the two sources are in the same cone of confusion.

Brimijoin et al. [33] reported that in the real world listeners are faced with acoustic environments that rarely consist of a single target sound and a single localizable masker. In real-life listening situations, we are confronted with multiple sound sources, either stationary or moving. Thus, conversations may become very difficult to understand with masking noises that have movements in space. Therefore, this experiment aims at quantifying the SRT and the SRM of a moving masker, understanding the role of listeners' head movements in a speech-in-noise task. For this purpose, a comparison between static and dynamic reproduction methods was made. In this way, it is possible to investigate if listeners use their head movements to maximize their intelligibility in an acoustic scene with a moving masker.

A virtual acoustic environment was simulated and binaurally reproduced to listeners in an attempt to address the following research questions: (1) Does dynamic binaural reproduction, by tracking listener head movement in the virtual scene to update the HRTF in real-time, benefit speech intelligibility and improve SRT and SRM?, (2) Does dynamic binaural reproduction differences in speech perception between masker moving in different directions (away from the target and toward the target)?, and (3) Are there differences in speech perception between stationary and moving masker when dynamic binaural reproduction is using?

For clarification, the terms "dynamic" and "static" are used to describe two methods of binaural reproduction with and without listener's head movement in the virtual acoustic scene, respectively; whereas "moving" and "stationary" are reserved for describing the masker trajectories.

8.1 Experimental Methodology

A total of 14 young adults (4 female, 10 male) participated in the task, aged between 20 and 27 years. All participants had normal hearing (pure tone thresholds

< 20 dB hearing level between 125 and 8000 Hz) at the time of the experiment and spoke German as their native language. Each participant was provided instructions regarding the tasks and they gave written consent prior to testing. The target stream of digit-triplet is defined in section 5.4. Speech stimuli samples were a set of German digit-triplet tests (see chapter 5). The masking noise was a randomized superposition of all digits used in the test (see subsection 5.1.5), resulted in a quasi-stationary noise with the same long-term averaged spectrum as the target speech.

For the audio playback, an auralization technique based on a binaural reproduction was used (see section 2.3). The virtual acoustic scenes were created using the software VA developed at the ITA [114] (see chapter 4). Positions of the listener and sound sources, as well as the moving trajectory of the masker, were defined in VA. The relative distance between the listener position and the positions of the target and masker was fixed at 1 m in all virtual acoustic scenes. The target was always located at $0°$ azimuth relative to the listener in VA. Depending on the specific test conditions, the masker source could be stationary at $0°$, $20°$, $45°$, and $70°$ azimuth or moving along an arc of $90°$, either toward or away from the target source position at an angular velocity of $32.7°/s$. There was always only one masker source in the virtual acoustic scene.

The same set of HRTFs measured from an artificial head developed at ITA [214], with a resolution of $3°$ in both azimuth and elevation angles, was utilized for binaural auralization to render virtual sound sources (see subsection 2.3.2). During the time-course of stimulus playback only the masker changes its position, hence producing a continuous but consistent change in both ITD and ILD during stimulus presentation as shown in Figure 8.1 and Figure 8.2 respectively.

Since in this thesis all stimuli were reproduced via headphones, the acoustic influence of headphones was considered. According to Masiero and Fels [166] an individual HpTF was measured and applied for all participants (see subsection 2.3.3). In half of the test conditions, when listeners' head movement was allowed in the virtual acoustic scenes, the binaural auralization was conducted in real-time during the experiment by updating the HRTFs based on the listener position in the virtual scene. An array of four optical tracking cameras in the listening booth was utilized to track a rigid body attached on a cap worn by the participant to capture the head movement at a sampling frequency of 120 frames per second and to update the listener position in VA (see chapter 4).

All trajectories involving spatial separation between the target and masker were tested with the masker oriented on the right quadrant of the median plane. To examine the effect of listeners' head movement, the seven core conditions (see Figure 8.3) varying in masker conditions were repeated in two blocks: with and without head tracker activated in the binaural reproduction. The 14 test

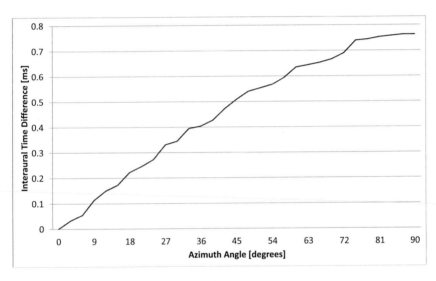

Figure 8.1: ITD as a function of azimuth angle and time on the condition masker moving 90°, with a 3° resolution. The ITD was calculated by cross-correlating the HRTFs after applying a bandpass filter with cutoff frequencies at 200 Hz and 1000 Hz, specifically for the useful frequency range from the speaker whose fundamental frequency was around 200 Hz.

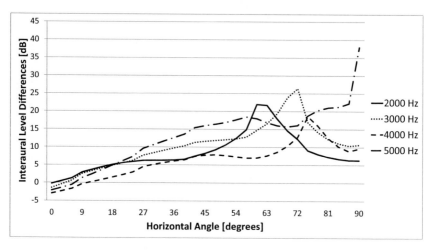

Figure 8.2: ILD as a function of azimuth angle and time on the condition masker moving 90°, plotted for 2000, 3000, 4000, and 5000 Hz. As the masker moves from 0° to 90°, ILD increases and reaches peak at different angles across the frequencies plotted.

conditions were arranged in a nested Latin Square, where the seven core conditions containing variations in masker conditions (stationary and moving), and trajectory (moving away and toward) were nested within the two test conditions of with versus without head movement (dynamic vs. static reproduction).

To investigate whether participants use their head movements to improve their intelligibility, the predictive model for speech intelligibility in noise of Jelfs et al. [119], which gives optimal orientation to the head for different locations of the target and the masker, was compared with participants' head movements. Figure 8.4 shows the predicted SRM as a function of head orientation for all stationary configurations. The initial directions of target and masker are indicated by subscripts (e.g., target at 0° and masker at 20° is denoted as $T_0 M_{20}$). The minimum values of SRM occur when the orientation of the head is such that both the target and the masker are in the same cone of confusion. The maximum values of SRM occur when the orientation of the head is: between $+30°$ and $+45°$ for $T_0 M_{90}$, $+30°$ for $T_0 M_{70}$, $-10°$ for $T_0 M_{45}$, and $-90°$ for $T_0 M_{20}$ (waiting for head movements not as long as 90°, the second optimal orientation is $-15°$).

The listening experiment took place in the ITA sound-attenuated listening booth. The room was completely in dark with both windows closed, this with the purpose that the listeners focus only on the acoustic stimuli (see section 4.2). Listeners gave their answers though loud voice and the evaluator, located outside the room,

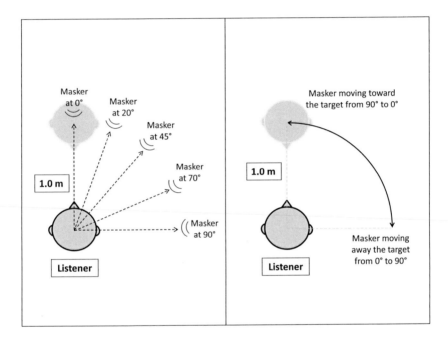

Figure 8.3: Stationary and moving masker conditions. Left side shows the spatial position of the masker for the five stationary cases. The right side shows cases with masker moving away from $0°$ to $90°$ and moving toward from $90°$ to $0°$.

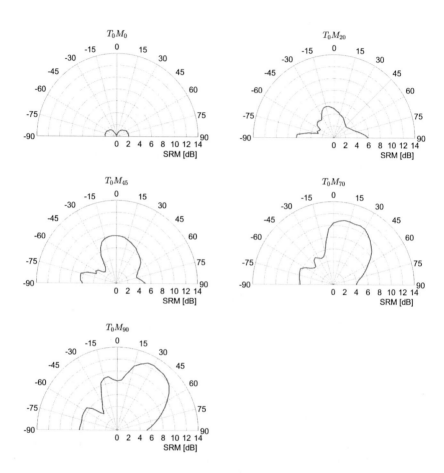

Figure 8.4: Predicted SRM in the five stationary spatial configurations: T_0M_0, T_0M_{20}, T_0M_{45}, T_0M_{70}, and T_0M_{90}.

entered the results using the keyboard. The instructions for the participants were: "Your task is to repeat back the digit-triplets heard in noise. You'll notice that sometimes the noise will move a little and the target voice may become progressively quieter. Do the best that you can to listen and repeat those digits back to me verbally".

Each digit in the triplet was scored correctly only when the digit itself and the sequential position were accurately identified. The possible score of each digit-triplet trial was 0 %, 33 %, 66 %, and 100 %. A digit-triplet was scored as a correct trial when $\geq$ 66 %. A simple up-down adaptive procedure [154] was used to track the SNR at 50 % speech intelligibility by changing the target speech level. The initial masker level was played back at 70 dB (re 20 μPa), resulting in 0 dB SNR. The initial step size was set at 8 dB until the first reversal was reached, from which the step size was 4 dB until the second reversal was reached, thereafter, the step size was set on 2 dB. The SRT was calculated using the MATLAB psignifit toolbox (version 3.0), applying the methods described by Wichmann and Hill [249, 250].

To examine whether there are differences between stationary and moving maskers, the performance of both moving masker conditions was predicted from the stationary data. By knowing the thresholds (SRTs) of the maskers at the three stationary conditions (20°, 45° and 70°), and the steepness of each psychometric function, it is possible to predict the SRT of a masker moving 90° (away or toward) combining the three stationary conditions at the same SNR level. Therefore, by comparing the SRT and the slope of the predicted psychometric function with the current moving results, it would be possible to know whether the masker's movement itself makes the task more difficult (or less) than a stationary condition. Figure 8.5 shows the psychometric functions of the three stationary masker conditions and the derived predicted psychometric function for a masker moving 90° (away or toward the target position). Figure 8.6 shows the psychometric functions of both moving masker conditions and the predicted psychometric function for a masker moving 90°.

8.2 Results

8.2.1 Speech reception threshold

To analyze the contribution of SRT, a two-way analysis of variance (ANOVA) was fitted to the factors: masker conditions (stationary masker at 0° vs. stationary

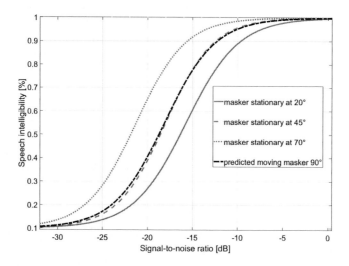

Figure 8.5: Psychometric functions of the three stationary masker conditions (20°, 45° and 70°) together with the predicted psychometric function of a masker moving 90° (away or toward the target position).

masker at 20° vs. stationary masker at 45° vs. stationary masker at 70° vs. stationary masker at 90° vs. moving masker away from the target on a 90° arc vs. moving masker toward the target on a 90° arc), and binaural reproduction (static vs. dynamic) as the two within-subject variables. Results show a significant main effect for masker conditions [$F(6,78) = 94.37$, $p < .001$], reflecting a trend of reduced masking effect as the target-masker separation enlarged. Pairwise comparisons with Bonferroni correction were used to examine the inter-relations between the masker conditions. Almost all possible pairs were statistically significant at $p < .05$, with masker stationary at 0° as the most effective interferer location. Non-significant differences ($p > .05$) were found in masker conditions pairs between (1) stationary masker at 45° and moving toward, (2) stationary masker at 45° and moving away, (3) stationary masker at 45° and stationary masker at 90°, (4) stationary masker at 70° and stationary masker at 90°, (5) stationary masker at 90° and moving toward, and (6) moving away and moving toward. Furthermore, a significant main effect for binaural reproduction [$F(1,13) = 15.08$, $p = .002$] was found, suggesting differences between with and without head movements. The interaction between masker conditions and binaural reproduction was not significant [$F(6,78) = 0.69$, $p < .05$], suggesting

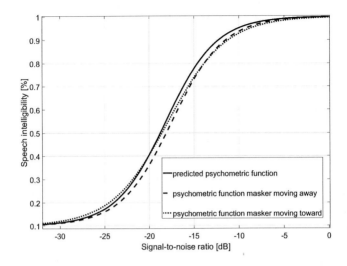

Figure 8.6: Psychometric functions of both moving away and toward masker conditions together with the predicted psychometric function of a masker moving 90°.

that dynamic reproduction does not alter the performance of speech perception under various masker conditions.

The relation between masker conditions and binaural reproduction is shown in Figure 8.7, representing mean SRTs measured under static and dynamic binaural reproduction for the seven masker conditions. Results using one-sample t-tests suggests that listeners achieved significantly higher SRT (worse intelligibility) in dynamic reproduction when the masker stayed at stationary conditions: (1) at $0°$ ($t(13) = $ -3.62, $p = $ .003), and (2) at $70°$ ($t(13) = $ -3.06, $p = $.009). However, for the two moving masker conditions there were no differences between binaural reproduction methods ($p > $ .05).

8.2.2 Spatial release from masking

A comparison of the results for the co-located condition with those for spatially separated conditions was analyzed in terms of the amount of SRM. To do so, the seven-level variable of masker conditions in the SRT results was further reduced to six levels in SRM: (1) stationary at $20°$, (2) stationary at $45°$, (3) stationary

81

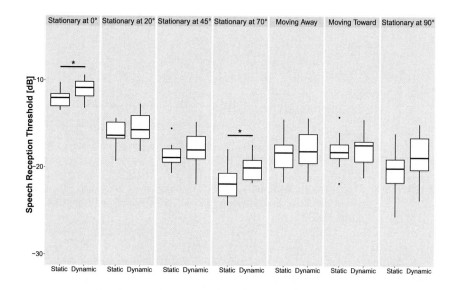

Figure 8.7: Speech reception thresholds at 50 % speech intelligibility measured for each masker condition: (a) stationary at 0°, (b) stationary at 20°, (c) stationary at 45°, (d) stationary at 70°, (e) moving away, (f) moving toward, and (g) stationary at 90°. In each masker condition, the SRT was plotted separately for with and without listener head movement tracked. Asterisks denote the significantly different pairs of SRTs measured with versus without head movement at $p < .05$. The error bars show 95 % confidence intervals. Asterisks denote the significantly different pairs of SRTs with *, $p < .05$ and **, $p < .001$.

at 70°, (4) moving away, (5) moving toward, and (6) stationary at 90°. A similar two-way ANOVA was fitted to the SRM data, again using masker conditions and binaural reproduction as the within-subject variables. A significant main effect was found for masker conditions $[F(5,65) = 29.66, p < .001]$, reflecting a trend of increasing SRMs as the target-masker separation enlarged. However, for binaural reproduction, the difference in SRM was not significant $[F(1,13) = 0.25, p > .05]$. The masker conditions x binaural reproduction interaction was not significant, $p > .05$. Figure 8.8 plots SRM for each masker condition separately for the two binaural reproduction methods: with and without listener head movement activated.

Bonferroni-corrected pairwise comparisons were applied to examine the main effect of masker conditions more closely. Results show no SRM differences between with and without head movements under all masker conditions ($p > .05$).

8.2.3 Stationary vs. moving masker

To compare between stationary and moving masker conditions under both dynamic and static binaural reproduction, a two-way ANOVA was fitted to the slope and SRT data. Thus, in the first analysis, the data was fitted with binaural reproduction (static vs. dynamic) and slope masker conditions (predicted slope vs. slope of masker moving away vs. slope of masker moving toward) as the within-subject variables. A significant difference in binaural reproduction was not found $[F(1,13) = 0.04, p > .05]$, but a significant difference for the slope masker conditions $[F(2,26) = 5.34, p = .011]$. Bonferroni-corrected pairwise comparisons were applied to examine this main effect more closely. A significant difference between the predicted slope and the slope of the masker moving away was found in the static reproduction ($p = .025$), but significant differences for the dynamic reproduction ($p > .05$) were not found. Figure 8.9 shows the slope in % against the three conditions, separately for static and dynamic binaural reproduction.

In the second analysis, the data was fitted with binaural reproduction (static vs. dynamic) and SRT conditions (predicted vs. moving away vs. moving toward) as the within-subject variables. A significant difference in binaural reproduction $[F(1,13) = 0.89, p > .05]$, and SRT conditions were not found $[F(2,26) = 0.19, p > .05]$. Figure 8.10 shows the SRT against the three conditions, separately for static and dynamic binaural reproduction.

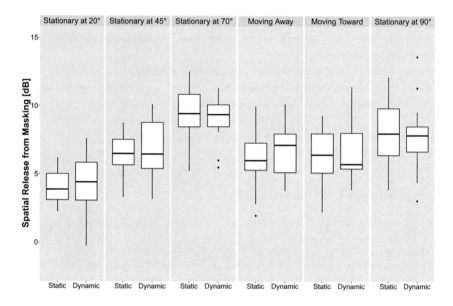

Figure 8.8: SRM measured for masker conditions: (1) stationary at 20°, (2) stationary at 45°, (3) stationary at 70°, (4) moving away, (5) moving toward, and (6) stationary at 90°. Mean SRMs measured in virtual acoustic environments with static binaural reproduction (listener head movement was not used to update head-related transfer functions, HRTFs) and with dynamic binaural reproduction (listener head movement was used to update HRTFs in real-time). The error bars show 95 % confidence intervals.

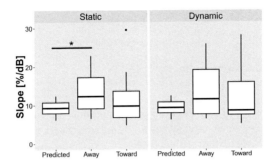

Figure 8.9: Slope measured for the predicted condition vs. slope of the masker moving away vs. slope of the masker moving toward. Mean slope measured in virtual acoustic environments with static and dynamic binaural reproduction. The error bars show 95 % confidence intervals. Asterisks denote the significantly different pairs of slopes with *, $p < .05$.

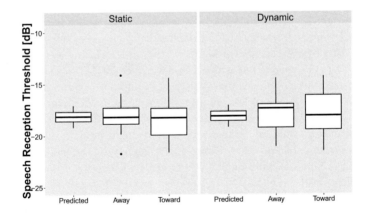

Figure 8.10: SRT measured for the predicted condition vs. SRT of the masker moving away vs. SRT of the masker moving toward. Mean SRT measured in virtual acoustic environments with static and dynamic binaural reproduction. The error bars show 95 % confidence intervals.

8.3 Discussion

8.3.1 Head movement behavior

Participants wore the head tracker cap throughout the entire experiment and they were unaware of when they were tested with head movement tracked to update HRTFs in real-time. All participants completed the task in one session. Grange and Culling [93] reported head movements larger or shorter than 10° separated from the target location. Therefore, in the present experiment also the 10° was used as a reference of spontaneous head movements. In the case stationary masker at 0°, 64 % of the listeners moved their head more than 10°, reaching 80° as the largest head movement. In the case stationary masker at 20°, 50 % of the listeners moved their head more than 10°, reaching 60° as the largest head movement. In the case stationary masker at 45°, 36 % of the listeners moved their head more than 10°, reaching 65° as the largest head movement. In the case stationary masker at 70°, 43 % of the listeners moved their head more than 10°, reaching 50° as the largest head movement. In the case stationary masker at 90°, 64 % of the listeners moved their head more than 10°, reaching 45° as the largest head movement. In the case moving masker away, 50 % of the listeners moved their head more than 10°, reaching 45° as the largest head movement. In the case moving masker toward, 50 % of the listeners moved their head more than 10°, reaching 54° as the largest head movement.

During the static binaural reproduction, the head tracking system was also recording the movements but without updating the HRTF in real-time. It was expected no head movements due to listeners could not perceive any change in the acoustic scene, nevertheless, on average of the seven conditions, the 36 % of the listeners move their head more than 10°. One of the possible reasons of these head movements could be because they were trying to update the HRTF as in the dynamic reproduction cases. However, the amount of head movements during the dynamic binaural reproduction was larger, on average 51 % of participants moved their head more than 10°. Therefore, more listeners moved their heads during the dynamic binaural reproduction due to the likelihood that they were achieving better intelligibility by turning their heads.

Grange and Culling [93], using real sound sources, reported that 9 of the 16 participants (56 %) moved their head more than 10° in a specific spatial separated case. In the current experiment, using virtual sound sources, 51 % of participants moved their head more than 10°. Therefore, the use of virtual sound sources seems a reliable option to evaluate the movements of the head, as well as real sound sources.

8.3.2 Stationary vs. moving masker comparison

The results only showed a difference between the predicted slope ($M = 9.3$ %/dB, $SD = 1.11$ %/dB) and the slope of the masker moving away ($M = 13.75$ %/dB, $SD = 3.27$ %/dB). It is important to mention that this difference was significant only in the static binaural reproduction. The comparison between predicted SRT and the SRT of the moving masker conditions were found not different for both binaural reproduction. Therefore, for a masker moving 90° away or toward the target position (0°), it is possible to predict its SRT from the psychometric functions of the stationary maskers.

8.3.3 Benefit of head movements on intelligibility

A significant difference in SRT was observed between the dynamic and static binaural reproduction in stationary masker conditions, but not in the moving masker conditions. Listeners performed higher SRTs (worse intelligibility) when their head movements updated the HRTFs in real-time by around 1-1.6 dB in the stationary masker conditions (Figure 8.7).

Results in this study show that dynamic reproduction had worse SRT in two stationary masker conditions, but not for the moving masker conditions. Therefore, listeners could have used their head movements more effectively to resolve the rapidly changing and ambiguous binaural cues while the masker was moving.

An evaluation of the head movements during the adaptive procedure was made to investigate if listeners improve the SNR during the task with the use of their head movements. The HRTF was updated to the next closest angle (in 3° resolution) when the listener moved their head to the left or to the right quadrant of the medial plane. For practical purposes, it was decided to show only three of the seven cases to exhibit the head movements: (1) masker stationary at 90° (as an example of all stationary cases), (2) masker moving away, and (3) masker moving toward. The upper part of Figure 8.11, Figure 8.12 and Figure 8.13 shows the head movements of the listeners in azimuth degrees, trial after trial as a unique timeline, meanwhile, the lower part shows the changes in SNR during the adaptive procedure. Therefore, it is possible to observe how the head movements are related to the changes in SNR, trial after trial, during each condition tested.

In the case stationary masker at $90°$, Figure 8.11 (a) shows the head movements of the two listeners who moved their head through the right quadrant of the median plane. Figure 8.11 (b) shows the head movements of the four listeners who moved their head through the left quadrant. Figure 8.11 (c) shows the head movements of the five listeners who moved their head through both quadrants. Knowing that the best orientation of the head is between $30°$ and $45°$ (to the right), it is possible to observe that the head movements of the participants do not follow any pattern towards that orientation. Even so, it is possible to observe the lowest SRT (better intelligibility) in the bi-directional movements (black line Figure 8.11 (c), lower part). This participant, during the first trials, oriented their head more to the left quadrant but when the task becomes harder (speech perception becomes harder trial after trial) the head movements were inclined more to the right quadrant reaching approximately $45°$ that is the optimal head orientation. By orienting their head in between the target/masker configuration, listeners are optimizing the binaural cues.

In the case moving masker away, Figure 8.12 (a) shows the head movements of the four listeners who moved their head through the left quadrant of the median plane. Figure 8.12 (b) shows the head movements of the three listeners who moved their head through both quadrants. In the case of the masker moving away, the optimal orientation of the head is expected to change in relation to the movement of the masker. Therefore the head must move from $-15°$ (best orientation $T_0 M_{20}$) to $+45°$ (best orientation $T_0 M_{90}$) during the stimulus. It is possible to observe the lowest SRT in the bi-directional movements (black line Figure 8.12 (b), lower part). Similar to the previous case, during the first trials the listener oriented their head more to the left quadrant but when the task becomes harder the head movements were more to the right quadrant, reaching almost $20°$. As was mentioned before, the most head movements were oriented only to the left, which could be because the movement of the masker. In this case, the masker was moving away from the target, thus, most of the listeners could have tried to put their right ear closer to the target position meanwhile the masker was separating.

In the case moving masker toward, Figure 8.13 (a) shows the head movements of the four listeners who moved their head through the right quadrant. Figure 8.13 (b) shows the head movements of the two listeners who moved their head through the left quadrant. Figure 8.13 (c) shows the head movements of the three listeners who moved their head through both quadrants. In the case of the masker moving toward, the head should change the orientation from $+45°$ (best orientation $T_0 M_{90}$) to $-15°$ (best orientation $T_0 M_{20}$). In head movements to the right is possible to observe the lowest SRT (black line Figure 8.13 (a), lower part). This participant oriented their head, almost during all trials, approximately at $20°$.

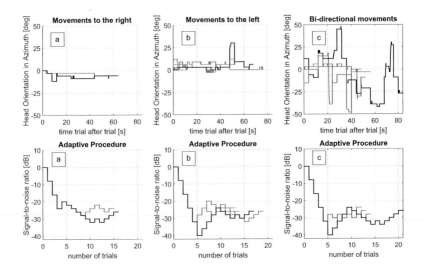

Figure 8.11: Head movements on the horizontal plane of the case masker stationary at 90°. (a) Show movements to the right and its corresponding
SNRs during the adaptive procedure to track the SNR at 50 %
speech intelligibility, (b) show movements to the left and its corresponding SNRs, and (c) show bi-directional movements and
its corresponding SNRs. The black line represents the lowest
SRT reached in each case and their correspondent head movements, meanwhile, the gray line represents the rest of listeners'
head movements and its corresponding SNRs during the adaptive
procedure.

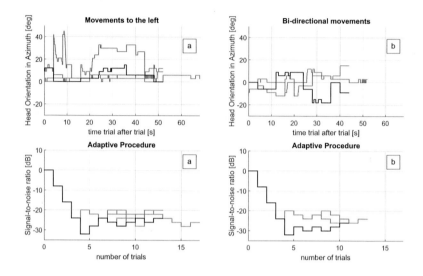

Figure 8.12: Head movements on the horizontal plane of the case moving masker away from the target. (a) Show movements to the left and its corresponding SNRs during the adaptive procedure to track the SNR at 50 % speech intelligibility, and (b) show bi-directional movements and its corresponding SNRs. The black line represents the lowest SRT reached in each case and their correspondent head movements, meanwhile, the gray line represents the rest of listeners' head movements and its corresponding SNRs during the adaptive procedure.

Unlike the case moving masker away, in this case, the head movements did not show an orientation trend. This could be due to the masker moving toward begins its movement separated 90° and finish at 0° co-located with the target, thus, the participants could have been more confused about where to orient their heads, maybe in between the target/masker (to the right) or away from the two sources (to the left).

During the adaptive procedure in each condition, where the SNR gradually varied, participants' head movement did incur in several short and large movements. However, the head movements did not seem to follow a clear pattern as testing progressed within each condition. In some cases, participants barely moved their heads, in others, moved their head from right to left in several consecutive trials resulting in no intelligibility benefit. In the most extreme cases, participants moved their head almost 80°; but such drastic head movements were not found beneficial for the speech perception. Therefore, head movements could provide intelligibility benefits but because of the erratically movements of the participants, all the potential benefits were not achieved.

The role of head movement in a speech-in-noise test was recently investigated by two studies using real sound sources that may shed light on results from this investigation. Both studies provided careful and unbiased instructions to listeners on their use of self-directed head movement in the experiments, similar to those provided in the current experiment. Brimijoin et al. [33] created specific conditions for asymmetrically hearing-impaired listeners where an optimal angle existed for maximum intelligibility through self-orientation. They found that self-directed head movements by listeners were highly variable, and some listeners self-oriented to a final angle away from the optimal angle by as much as 100°. A study by Grange and Culling [93] used normal-hearing listeners and provided even more rigorous control of the experimental protocol by tracking the free head movement throughout the experiment. Their results suggested irrational head movement among the 16 listeners tested and no consistent regularity could be reported from the patterns observed. Similar to findings in the current experiment, both studies seemed to suggest that listeners do not spontaneously find intelligibility improvement from self-directed head movement in a speech-in-noise task. Neither study, however, compared the performances with those measured with fixed head orientation in the experiments.

One of the possible reasons why the reproduction method was found non-significant could be the use of an artificial head HRTF, instead of individual HRTFs. It is known that the use of individual HRTFs can provide a better "immersion" in the virtual scene [237], therefore, could be possible to observe more intentional head movements.

8.4 The Role of Individual HRTF

In subsection 2.3.4 it was established that the use of mismatched HRTF data-sets, as artificial head HRTFs, can result in unwanted artifacts in coloration and localization. Therefore, the results in the previous experiment could have been affected by this fact.

The use of individual HRTFs has been studied in several tasks such as sound localization (error rate), externalization of virtual sound sources, and front-back

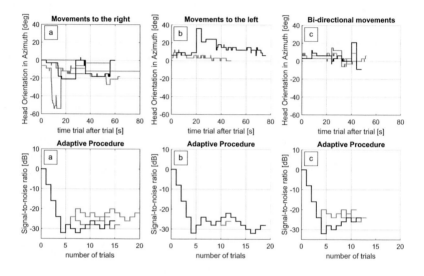

Figure 8.13: Head movements on the horizontal plane of the case moving masker toward to the target. (a) Show movements to the right and its corresponding SNRs during the adaptive procedure to track the SNR at 50 % speech intelligibility, (b) show movements to the left and its corresponding SNRs, and (c) shows bi-directional movements and its corresponding SNRs. The black line represents the lowest SRT reached in each case and their correspondent head movements, meanwhile, the gray line represents the rest of listeners' head movements and its corresponding SNRs during the adaptive procedure.

confusion. Morimoto and Ando [175] found that the use of individual HRTFs provides better accuracy of sound localization. But nevertheless the findings of Morimoto and Ando, Begault et al. [14], using an individualize HRTF, did not find significant differences in externalization, front-back confusion, and localization tasks. After that, Oberem et al. [184, 183] found only slight differences in front-back confusion and localization tasks, concluding that the use of non-individual HRTFs could be a less complex and elaborate way to successfully provide for a realistic auditory perception.

Despite all previous studies, little is known about the role of individual HRTFs in speech-in-noise tasks when conducted in a virtual acoustic environment (VAE). Therefore, individual HRTFs and head movements were assessed to better understand their roles in facilitating speech-in-noise perception. At the same time, differences in speech perception between stationary and moving masker when is using individual HRTFs and dynamic binaural reproduction were evaluated.

8.4.1 Procedure

A total of 11 young adults (6 female) participated in the task, aged between 20 and 26 years. All participants had normal hearing (pure tone thresholds < 20 dB hearing level between 125 and 8000 Hz) at the time of the experiment and they spoke German as their native language. Each participant was provided instructions regarding the tasks and they gave written consent prior to testing. Each listener participated in a previous session to measure their individual HRTF (subsection 2.3.4).

The target stream of digit-triplet and masker noise was previously defined in chapter 5. The listening experiment took place in the same sound attenuated listening booth previously mentioned. Again, no display in front of participants was used to prevent their attention from focusing on the screen. Listeners gave their answers though loud voice and the evaluator, located outside the room, entered the results using the keyboard. The same instructions presented in Grange and Culling [93] were offered to each participant.

For the audio playback, an auralization technique (see section 2.3) based on a binaural reproduction was used. The virtual acoustic scenes were created using VA. Positions of the listener and sound sources, as well as the moving trajectory of the masker, were defined in VA. The relative distance between the listener position and the positions of the target and masker sound sources was fixed at 1 m in all virtual acoustic scenes. The target was always located at 0° azimuth

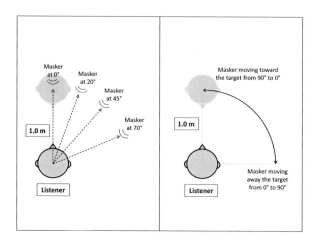

Figure 8.14: Spatial location of masker assessed in the experiment. The M represents the position or movement of the masker.

relative to the listener in VA. Depending on the specific test conditions, the masker could be stationary at 0° and 90° azimuth or moving from 0° to 90° away from the target. There was always only one masker source in the virtual acoustic scene. All these conditions are shown in Figure 8.14.

The same set of HRTFs measured from an artificial head developed at ITA [214] with a resolution of 1° in both azimuth and elevation angles, was utilized for binaural auralization to render virtual sound sources (see subsection 2.3.2).

In half of the test conditions, when listeners' head movement was allowed in the VAE, the binaural auralization was conducted in real-time during the experiment by updating the HRTFs based on the listener position in the virtual scene. An array of four optical tracking cameras was utilized to track a rigid body attached on a cap worn by the participant to capture the head movement (see chapter 4). To ensure that participants were not conscious of changes in the head tracking condition, they were asked to wear the head tracker cap during the whole test session.

To examine the effect of listeners' head movement, the three core masker conditions were tested, with and without head tracker activated in the binaural reproduction. The 12 test conditions were arranged in a nested Latin Square, where the three core masker conditions (stationary vs. moving), type of HRTF used (artificial head HRTF vs. individual HRTF), and binaural reproduction (static vs. dynamic), were nested.

Each digit in the triplet was scored correctly only when the digit itself and the sequential position were accurately identified. The possible score of each digit-triplet trial was 0 %, 33 %, 66 %, and 100 %. A digit-triplet was scored as a correct trial when $\geq$ 66 %. A simple up-one adaptive procedure [154] was used to track the SNR at 50 % speech intelligibility by changing the target level. The initial distractor noise level was played back at 70 dB (re 20 μPa), resulting in 0 dB SNR. The initial step size was set at 4 dB until the first reversal was reached, from which the step size was 2 dB thereafter. The SRT was calculated using the MATLAB psignifit toolbox (version 3.0), applying the methods described by Wichmann and Hill [249, 250].

8.4.2 Results

Speech Reception Threshold

The SRT data was analyzed in a repeated-measures analysis of variance (ANOVA) with masker conditions (stationary at $0°$ vs. moving away $90°$ vs. stationary at $90°$), HRTF (artificial head HRTF vs. individual HRTF) and binaural reproduction (static vs. dynamic) as the within-subject variables.

Results show a significant main effect for masker conditions [$F(2,20) = 135.14$, $p < .001$], reflecting a trend in which stationary at $0°$ was the most effective masker location and stationary at $90°$ the lowest SRTs. Pairwise comparisons with Bonferroni correction were used to examine the inter-relations between masker conditions, revealing that all possible pairs were statistically significant at $p < .001$. There is also a significant main effect in HRTF [$F(1,10) = 8.96$, $p = .013$]. Pairwise comparisons with Bonferroni correction were used to examine the mean difference more closely. In the masker condition stationary at $90°$ the HRTF was found significant different ($p = .008$), but non-significant differences in the cases of masker stationary at $0°$ and moving $90°$ ($p > .05$). The binaural reproduction was found non-significant [$F(1,10) = 1.07$, $p > .05$], suggesting that there are no differences between with and without head movements. All interactions were found to be statistically non-significant ($p > .05$).

Mean SRTs for the three masker conditions are shown in Figure 8.15, separated for static and dynamic reproduction, and artificial head and individual HRTF.

Spatial Release from Masking

A comparison of the results for the co-located condition with those for spatially separated conditions was analyzed in terms of the amount of SRM. A similar ANOVA analysis was fitted to the SRM data, again using masker conditions

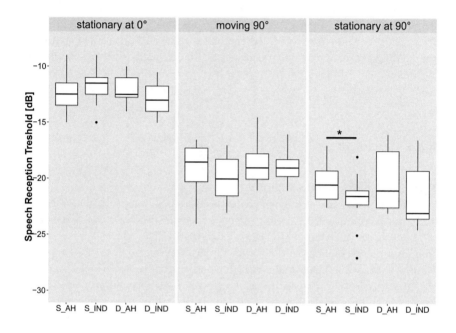

Figure 8.15: Speech reception thresholds at 50 % speech intelligibility measured for each masker condition: (a) stationary at 0°, (b) moving 90°, and (c) stationary at 90°. In each masker condition, the SRT was plotted separately for both HRTF used and for both binaural reproduction methods. S_AH represent static reproduction and the used of artificial head HRTF. S_IND represent static reproduction and the used of individual HRTF. D_AH represent dynamic reproduction and the used of artificial head HRTF. D_IND represent dynamic reproduction and the used of individual HRTF. The error bars show 95 % confidence intervals. Asterisks denote the significantly different pairs of SRTs with *, $p < .05$.

(moving 90° vs. stationary at 90°), HRTF (artificial head HRTF vs. individual HRTF), and binaural reproduction (static vs. dynamic) as the within-subject variables. A significant main effect was found for masker conditions $[F_{(1,10)} = 11.41, p = .007]$, reflecting a tendency of lower SRMs in the masker moving 90° than the masker stationary at 90°. For binaural reproduction a non-significant main effect was found $[F_{(1,10)} = 3.48, p > .05]$. Also, for HRTF a non-significant main effect was found $[F_{(1,10)} = 2.85, p > .05]$. All interactions were found to be statistically non-significant $(p > .05)$. Figure 8.16 plots SRM against masker conditions separately for static and dynamic reproduction, and HRTFs used in the experiment.

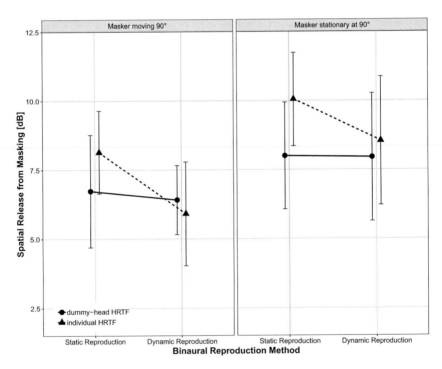

Figure 8.16: Spatial release from masking (SRM) measured for masker moving 90°, and stationary at 90°. Mean SRMs measured in virtual acoustic environments with static and dynamic binaural reproduction, and for both HRTFs used in the binaural synthesis (artificial head HRTF and individual HRTF). The error bars show 95 % confidence intervals.

Masker Conditions Analysis

In addition, a t-test analysis was made to compare between stationary and moving masker. To do so, the SRMs of moving masker 90° were compared to the accumulated SRM using the Jones and Litovsky SRM model (see chapter 7). The accumulated SRM, using the mathematical model, was 6.38 dB. A significant difference was found at level $p = .011$, between the accumulated SRM and the masker moving 90° for the static binaural reproduction and using individual HRTFs ($M = 8.13$ dB, $SD = 1.4$ dB). All other three cases were found non-significant different ($p > .05$).

8.5 Discussion and Conclusions

Individual HRTF

A significant main difference was found on SRTs in the type of HRTF used. A better listener's intelligibility was expected for the use of individual HRTFs than artificial head HRTF. Nevertheless, the significant difference was found only in the case when masker was stationary at 90°. It is known that better intelligibility is achieved when the masker is spatially separated and, in this case, the masker stationary at 90° had the larger advantage. Thus, a possible explanation is that the use of individual HRTFs could have a significant advantage in a speech-in-noise task only in cases with large spatial separation between target and masker, generating interdependence between these two factors that should be analyzed more in-depth in future works. Nevertheless, the difference was found only with the use of the static reproduction method (no head movements).

It was established that individual HRTFs bring a little benefits in localization tasks [184, 183, 14]. Thus, the current findings are in accordance with the previous studies since, for speech-in-noise perception tasks, the use of individual HRTFs also seems to provide only slightly benefits.

Binaural Reproduction method

In the first experiment, a significant difference only in two stationary cases was found. Nevertheless, the results in the second experiment show a non-significant SRT differences in the binaural reproduction. With the use of individual HRTF, more rational head movements were expected due to the higher resolution in the perception of the acoustic scene. Nevertheless, unlike what was expected, the use of individual HRTF with a dynamic reproduction did not improve the speech-in-noise perception. It is necessary to remember that this thesis tried to assess intelligibility when only acoustic cues were presented because no visual cues were offered to listeners. Insisting on the idea that head movements do have an influence in speech perception in real-life situations, it could be concluded that the use of only acoustic cues is not sufficient to assess real-life head movements and visual cues must be presented to listeners.

The previous SRM results showed no differences in the binaural reproduction, nevertheless, a significant difference was expected with the use of individual HRTFs. However, a non-significant difference was found between binaural reproduction methods. It is only possible to observe a tendency of lower SRMs, for both masker conditions, in the dynamic reproduction when individual HRTF was used. It is possible to suggest that, in a more realistic environment taking into account the natural irrational head movements, there are lower SRM benefits.

Comparison Between Stationary and Moving Masker

By means of the same strategy used in chapter 7, the masker moving results were compared with the accumulated SRM calculated with the Jones and Litovsky SRM model [121]. The calculation of the accumulated SRM using the predictive model was 6.3 dB. A one-sample t-test show a significant difference between the accumulated SRM and the masker moving ($M = 8.1$ dB, $SD = 1.4$ dB) for static reproduction and individual HRTF ($p = .019$). All other three combinations were not significant ($p > .05$). Thus, the effect due to the movement of the masker for static reproduction using individual HRTF was around 1.8 dB, increasing the SRM. Therefore, it is possible to assume that the use of individual HRTFs helps in the perception of moving sound sources better than for stationary sources.

9

Assessment of Different Reverberant Conditions in Young and Elderly Subjects at Circular and Radial Masker Conditions

In real-world listening situations, we often listen to speech in the presence of masking noises inside rooms with different reverberant conditions. Likewise, most everyday listening situations consist of multiple sound sources both stationary and moving, creating multiple acoustic reflections that reach our ears at different time and intensity [124]. Acoustic reflections of a speech signal can sometimes be beneficial because can increase the signal energy reaching the listener, in comparison with the anechoic condition, where the signal energy is largely absorbed by nearby surfaces. Nevertheless, reflections could also be harmful, because can superimpose on the direct sound, altering the waveform of the speech. Accordingly, many experiments have shown that reverberant conditions interfere with the speech-in-noise perception, becoming worse if reverberation increases (see section 3.5). One of the main reason for the reduction of speech perception is because the acoustic reflections reach the ears at different time and intensity, thus, interfering with the binaural advantage by reducing the extent to which the head effectively attenuates high-frequency sounds and by disrupting the fine timing cues used in the binaural analysis [124]. The detrimental effect of reverberation can also be linked to the decorrelation between the signals arriving at the two ears. The interaural cross-correlation (IACC) coefficient of a binaural signal describes the similarity of the incident sound waves at the two ears of the listener [150]. The IACC in a room is typically decreased by higher reverberation [103]. As a result, speech-in-noise perception diminishes when the IACC of the masker is decreased [147, 148, 156].

At the same time, room acoustics can notably reduce the speech perception for elderly subjects. Room reflections have a smoothing effect on the waveform of the signal, thus creating a distortion of the temporal waveform affecting elderly listeners to a greater extent [92]. It was shown that elderly listeners with normal hearing, minimal hearing loss, or severe hearing loss, achieve poorer speech perception than young listeners, even with comparable hearing sensitivity [92, 164]. This also extends to everyday conversation environments, where elderly subjects have more difficulties understanding speech than their younger counterparts do. Indeed, many studies reported that elderly subjects have difficulties following one-on-one conversations in noisy environments, besides the problem is exacerbated when they have to follow two or more talkers because they miss part of the speech content or have a lack of confidence in the part that they heard. Thus, elderly subjects are prone to anxiety or frustration and may avoid or be excluded from social interactions. Consequently, it is not surprising that elderly subjects consider speech perception declines as one of the most serious consequences of the aging process [176]. The research evidence, supporting age-related deficits for speech-in-noise perception, is mixed and appears to depend on a number of variables including the audibility of the speech signal, the type of speech signal (letters, words or sentences), the type of noise background (steady-state, modulated noise, or speech), the signal-to-noise ratio (SNR), and subject variables, including efforts to equate the hearing thresholds between younger and elderly subjects [92]. A well-established theory in aging research suggests that a generalized slowing in brain function is responsible for most, if not all, the age-related declines in problem-solving, reasoning, memory, and language [216]. According to this theory, slowing in brain functioning is thought to reduce the speed at which various cognitive operations can be performed. For example, it is generally assumed that the reason why elderly subjects often find it difficult to understand someone who is talking rapidly or fail to follow a conversation when there are multiple speakers, is because the rate of flow of information approaches or exceeds the maximum rate that can be accommodated by the cognitive processes involved in language comprehension [216]. However, when young and elderly subjects perform a task, including perceptual tasks at the same level of proficiency, mounting evidence indicates that different areas of the brain are activated depending on the age of the person, with a general pattern indicating that elderly subjects use more brain regions including regions of both hemispheres [199]. Therefore, it may be particularly difficult to evaluate the relative contribution of cognitive-level effects to speech-processing declines because, in such situations, both sensory and cognitive factors could be responsible for the speech-processing declines [155]. Research has shown that deterioration due to age occurs on many fronts such as hearing sensitivity declines, dynamic range is reduced, speech-in-noise perception is compromised, and cognitive processing slows, to name just a few. Good communication in complex listening environments requires the peripheral auditory systems, central auditory pathways, and cognitive systems to all function effectively. If the process is impeded at any one point, then, the ability to recognize speech breaks down [89].

In daily listening environments, maskers are not only separated angularly, but also in distance. Most of the researchers have studied speech perception only for stationary maskers located at circular horizontal locations [124, 169, 260] and very few studies have looked into the speech-in-noise perception of maskers at different radial distances [41, 221, 247]. Also for localization tasks, the distance perception is an aspect which has been given considerably less attention than horizontal localization. For auditory distance perception, several cues are available [263], where the most predominant cue is the signal intensity [229]. As distance increases, signal levels for omnidirectional sound sources decrease proportionally to the inverse of their distance [142]. Shinn-Cunningham et al. [221], and Brungart and Simpson [41] investigated spatial release from masking (SRM) related to differences in distance combined with angular separation. Both studies found a substantial effect of distance on SRM, however, they focused on interaural level difference (ILD) cues at very close distances (< 1 m) and only tooking into account anechoic environments. Meanwhile, Bronkhorst and Plomp [36] included reverberant conditions and considered sources at near and far distances. Their results showed an improvement in speech reception threshold (SRT) of $\sim$1 dB when moving the masker from near to far distance while keeping the target near. For its part Marrone et al. [164] determined that the effect of distance on SRM is mainly a monaural process and the binaural benefit is limited to $\sim$3 dB. Westermann and Buchholz [247] varied the maskers in distances of 0.5 m, 2 m, 5 m and 10 m, meanwhile, the target was fixed at 0.5 m in a room acoustic condition. Their results showed SRM benefits of about 10 dB for a speech masker at 10 m distance, but no SRM benefits for a speech-modulated noise masker at 10 m distance.

Therefore, this experiment investigates the effect of room acoustic in a speech-in-noise task performed by young and elderly subjects with a circularly or radially masker trajectories, manipulated in stationary or moving conditions. It also deals with the question if there are differences in speech-in-noise perception between stationary and moving maskers in different room acoustics conditions. We expected lower speech perception subject to moving maskers since the combination of room acoustic effects and the movement of the masker could represents a more complex listening situation, especially for elderly subjects. The aims of this thesis are (1) to investigate speech-in-noise perception of a moving masker under different room acoustic conditions, assessing two different masker movements: circular and radial, (2) compare SRT and SRM between stationary and moving masker, and (3) examine SRT and SRM for young and elderly subjects to evaluate between-group differences.

9.1 Experimental Methodology

A total of 24 participants took part in this experiment, all of them German native speakers. A cognitive test to identify participants with mild cognitive impairment was carried out before beginning the experiment. The *DemTect* test was chosen because of its short duration (8-10 min), and as it can be applied to subjects independent of their age and education [122]. All participants passed the cognitive test. The 24 participants were recruited to form different groups based on their age. The 12 young adults (4 female), aged between 21 and 29 years ($M = 24.3$, $SD = 2.5$), had normal hearing (< 20 dB hearing level, HL), determined by pure-tone audiometry (PTA) with test frequencies between 125 Hz and 8000 Hz. The 12 elderly participants (8 female), aged between 60 and 78 years ($M = 69.5$, $SD = 5.5$), were required to have PTA thresholds ≤ 25 dB HL between 0.25 and 3 kHz (ANSI S3.6 –1989) in both ears. Exceptions to this rule are allowed if a subject had a hearing level > 25 but ≤ 35 dB HL at only one frequency. There were 3 participants who fell into this category. Although audiograms within these thresholds are typical for elderly people whose hearing is considered to be clinically normal [216], their hearing is not equivalent to that of younger adults. None of the elderly participants used a hearing aid.

9.1.1 Virtual stimuli

To investigate speech perception for different room acoustic conditions, a room simulation was applied in this thesis, making it possible to specify source and receiver positions and control room acoustical parameters such as the reverberation time. The simulation environment RAVEN (see section 4.1) was chosen to create binaural room impulse responses (BRIRs) for different sound source positions in a virtual room. RAVEN applies a hybrid approach combining an image source model for early reflections [5] and a ray tracing algorithm for late reverberation [141]. In addition to the room geometry, frequency-dependent absorption and scattering coefficients of the virtual room's surface materials were included when running a simulation. The characteristics of a sound source can be described by a source directivity, containing one-third octave magnitude spectra measured on a spherical grid. The binaural receiver of the scene is characterized by a set of head-related transfer functions (HRTFs), for example, measured from an artificial head. Binaural characteristics were applied to all parts of the BRIR, namely, direct sound, early reflections and to the reverberation tail.

All RAVEN results applied in this thesis were simulated in octave-band resolution for 20,000 raytracing particles and a temporal histogram resolution of 10 ms. For the early part of the room impulse response (RIR), an image source order of 2 was chosen. All simulated BRIRs had a length of 2 s and a sampling rate of 44100 Hz.

This thesis investigates a virtual scene, consisting of a listener, a target source and masker for three different environments, called *anechoic*, *treated* and *untreated* in the following. While for the *anechoic* condition, only binaural direct sound synthesis without any reflections was calculated, a shoebox room with the dimensions 7 m x 6 m x 2.8 m^3 (room volume V= 117.6 m^3; surface area S = 156.8 m^2) for *treated* and *untreated* conditions was simulated (see Figure 9.1). Absorption and scattering coefficients were applied homogeneously to all six surfaces of the room. While the scattering coefficients had a frequency-independent value of 0.1, the absorption coefficients were frequency-dependent ranging from 0.21 to 0.30 and 0.06 to 0.15 in *treated* and *untreated* conditions, respectively. The use of different absorption coefficients led to different reverberation times T_{20} of 0.5 s (averaged for 500 Hz and 1 kHz frequency band) for the *treated* and 1.31 s for the *untreated* condition. The evaluation of reverberation times (RT) was done according to ISO 3382-1 [1], using the *ita_roomacoustics* function [43] of the ITA-Toolbox [18], an open-source project for Matlab (The MathWorks Inc., Natick, MA). For this, in total 12 RIRs were simulated for an omnidirectional sound source at two different source positions and six omnidirectional receiver positions in the room.

The frequency-dependent evaluation of averaged T_{20} values is shown in Figure 9.2. Values for the *untreated* condition were higher for all evaluated octave frequency bands than for the *treated* condition. While the high reverberation times above 1 s of the *untreated* condition could correspond to a completely empty room without any absorption material, the *treated* condition rather represents a fully furnished room with absorptive materials, such as carpets and curtains.

For the simulation of the BRIRs, positions of the listener, the target sound source and the stationary maskers, as well as the trajectory of the moving maskers, were defined in the simulation environment. The target sound source was always located at 0° azimuth relative to the listener, at a distance of 1.5 m. To account for the directionality of the human voice, a source directivity dataset was applied to the sound source [133]. For the receiver, an HRTF dataset of the artificial head, developed at the Institute of Technical Acoustics (ITA) Aachen [214], with a resolution of 3 degrees in both azimuth and elevation angles, was assigned.

Each HRTF had a length of 256 samples, using a sampling rate of 44100 Hz.

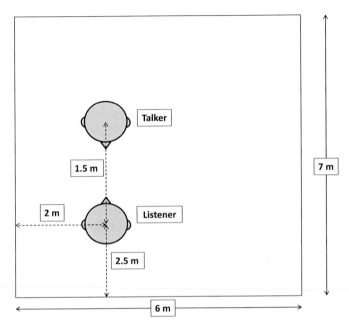

Figure 9.1: Spatial configuration of the listener and talker in the virtual room.

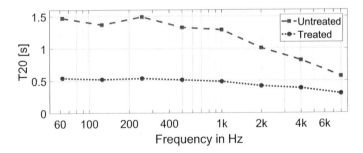

Figure 9.2: Simulated reverberation times T_{20} for the room conditions *treated* and *untreated*. Values were averaged for 12 evaluated room impulse responses including six receiver positions and two sound source positions.

The simulated BRIR files of all three room conditions were then convolved with the anechoic sound files of the experiment. Speech target stimuli samples were an extended German digit-triplet test (see chapter 5). The masker noise was a randomized superposition of all digits used in the test, resulted in a quasi-stationary noise with the same long-term averaged spectrum as the target speech (see subsection 5.1.5).

Except for the condition where the masker was at $0°$ azimuth, all other masker conditions involved spatial separation and were tested with the masker with locations in the right front quadrant with respect to the listener. Having one masker to the side of the listener represents an asymmetric masker array, thus, the component that provides larger SRM benefits is the head shadow effect [90, 121].

The examined conditions in this experiment were divided into two blocks: *circular* and *radial* conditions. There was always only one masker source in the virtual acoustic scene. The 24 test conditions were arranged as nested Latin Square, where the eight core cases containing variations of Masker Condition (*stationary* vs. *moving*), were crossed with the three Reverberation (*anechoic* vs. *treated* vs. *untreated*).

For the *circular* conditions, the masker was either stationary at discrete positions $0°$ (co-located with the target), $20°$, $45°$, or $70°$ azimuth, or performed a circular movement from $0°$ to $90°$ azimuth, representing the condition *circular moving masker*. In all cases, the distance of 1.5 m between the listener position and the masker is kept unchanged (see Figure 9.3).

It was considered that a direct comparison between conditions with a stationary masker and circularly moving masker is not accurate, and thus, the three $20°$, $45°$ and $70°$ stationary positions were considered jointly in the SRT and SRM evaluation, resulting in the same geometrical average position as when evaluating the circular moving masker condition. Figure 9.4 shows the angular positions of the moving masker conditions when playing back each of the three digits. When the masker moved circularly, the midpoint of the first digit was played back when the masker was at $20° \pm 2°$, the midpoint of the second digit was played back when the masker was at $45° \pm 2°$, and the midpoint of the third digit was played back when the masker was at $70° \pm 2°$. For that reason, the stationary masker that was compared with the *circular moving masker* was the average SRT value at $20°$, $45°$ and $70°$, and is referred to as condition *stationary circular masker*.

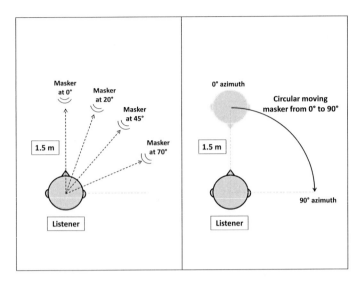

Figure 9.3: Stationary and moving maskers in the *circular* condition. The left side shows the four angular positions of the masker on the stationary conditions. The right side shows the condition with *circular moving masker* (from 0° to 90° azimuth).

For the *radial* conditions, the spatially separated masker was located at 70° azimuth but varied in distance. In addition to the co-located masker at 0° at a distance of 1.5 m, the stationary masker positions at 70° azimuth were defined at distances of 0.8 m, 1.15 m and 1.5 m. The conditions with the masker located at 0° and 70° at a distance of 1.5 m are the same as those used in the *circular* conditions. In the moving masker condition, the masker performed a linear movement from a distance of 0.5 m to 1.8 m and is referred to as *radial moving masker*. This movement corresponded to a reduction in sound pressure level (SPL) of $\Delta L = 11.1$ dB for the *anechoic* condition, according to the distance law.

Similar to conditions with circular masker manipulation, three digits were played back when the masker was at different distances in the *radial moving masker* condition. The three stationary cases shown in Figure 9.5 correspond to those different distances. In this way, the stationary masker that was compared with the *radial moving masker* was the average value of the SRTs at 0.8 m, 1.15 m, and 1.5 m, representing the *stationary radial masker*. For a better clarification, the different masker conditions are shown in Table 9.1.

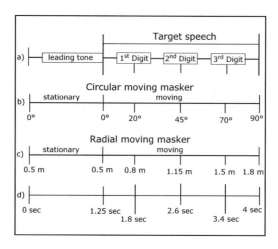

Figure 9.4: Illustrations of time events during playback of both target and
masker streams. In the virtual scene, masker movements start after
the leading tone. (a) Illustrations of the digit-triplet stimulus playback stream. (b) Mean angular positions of the continuous circular
masker movement, indicating time instances of individual digit playback. (c) Mean radial positions of the continuous radial masker
movement indicating time instances of individual digit playback.
(d) Timeline for stimulus streams.

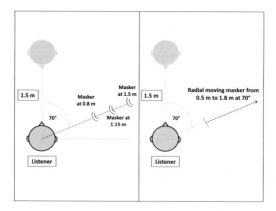

Figure 9.5: Stationary and moving maskers in the *radial* condition. The left
side shows the three radial positions of the masker on the stationary
conditions. The right side shows the condition with *radial moving
masker* (from 0.5 m to 1.8 m at 70° azimuth).

Table 9.1: Stationary and moving masker definitions for *circular* and *radial* conditions.

Examined conditions	Factor	Factor levels	Refered to as	Note
Circular	Masker Conditions	*stationary*	*stationary circular masker*	average SRT value at 20°, 45° and 70°
		moving	*circular moving masker*	
Radial	Masker Conditions	*stationary*	*stationary radial masker*	average SRT value at 0.8, 1.15 and 1.5 m
		moving	*radial moving masker*	

The binaural signals for the moving maskers are based on 91 BRIRs using an angular increment of $\Delta\theta = 1°$ and on 65 BRIRs using an increment of $\Delta r = 2$ cm for the circular and the radial movement, respectively. To generate the signals, time frames of 1,536 samples length ($f_s = 44.1$ kHz) of the anechoic signal were convolved with the corresponding BRIR. To guarantee a smooth crossover in case of changing BRIRs, a two-sided Hann window with a length of 512 samples was applied to each frame, overlapping 256 samples with the previous and next time frame.

To quantify the binaural coherence of the simulated BRIRs, the interaural cross correlation (IACC) coefficient was calculated for all positions of the masker. This was done using the function *ita_roomacoustics_IACC* of the ITA-Toolbox [18], which calculates the IACC according to ISO 3382-1. Prior to the calculation of the broadband IACC for the early part of the binaural room impulse responses (time interval $t_1 = 0$ s, $t_2 = 0.08$ s), a 10-th order Butterworth bandpass filter with cutoff frequencies $f_1 = 60$ Hz and $f_2 = 6$ kHz was applied. The bandpass frequencies were chosen according to the frequency spectrum of the digits and the masker noise, which show no substantial spectral components below 60 Hz and above 6 kHz.

Since in this experiment the auralized signals were presented to the listeners through a pair of headphones (HPs, Sennheiser HD 650), the acoustic influence of the HPs' transducers had to be compensated in order to obtain an accurate binaural reproduction. Therefore, headphone transfer functions (HpTFs) were measured individually for each test subject (see subsection 2.3.3).

9.1.2 Apparatus and procedure

The listening experiment took place in a sound-attenuated hearing booth at ITA which has a room volume of V $\approx$ 10.5 m^3 (L x W x H $[m^3]$ = 2.3 x 2.3 x 2.0). Participants were seated in front of a display and used a keyboard for data input. A GUI and test routine were developed in Matlab to playback test stimuli, record and evaluate responses, and perform the adaptive adjustment of SNRs. Each participant was provided instructions regarding the tasks and they gave written consent prior to testing. The listener's task consisted of identifying the three digits of each trial. Once both the target and masker streams finished, participants were able to enter the digit triplet in the text input field of a graphical user interface (GUI) using a keyboard. When the subject was not able to identify one or more digits, the digits could either be guessed or the corresponding input field was just left blank.

Each digit in the triplet was scored as correct only when the digit itself and its sequential position were both correctly identified by the subject. Possible scores per trial were 0 %, 33 %, 66 %, and 100 %, where a digit triplet was scored as correct trial for a score of 100 %. The adaptive up-down procedure proposed by Plomp and Mimpen [202] was used to track the SNR at 50 % speech intelligibility by changing the target speech level while keeping the masker level constant at 65 dB during the experiment. The initial SNR level was set to a value of -32 dB. If participants provided a wrong answer during the first trial, the same stimulus was repeated with the target level increased by 4 dB. After the first reversal, the step size was decreased, reducing or increasing the target level by 2 dB, depending on the subject answer. Each test condition finished after participants had reached eight reversals in total. The SRT was then calculated by taking the mean SNR values at the last eight reversals.

To measure an effective SNR, a speech-weighted SNR was calculated for each masker condition [97]. The speech-weighted SNR takes into account the relative contribution of different frequencies to speech intelligibility, and thus, providing a more effective measure [243]. The speech-weighted SNR was obtained by bandpass-filtering both signals (target and masker) in third-octave bands with center frequencies between 160 Hz and 8000 Hz and weighting them according to their contribution to speech intelligibility, as indicated in Table III of the Speech Intelligibility Index standard (ANSI, 1997). For calibration purposes, the root mean square (RMS) value of the binaural target and masker signals were adjusted to obtain the desired SNR in each trial of the adaptive procedure. At the same time, all signals were calibrated such that the signal at the right ear corresponded to the desired level as the masker was always presented to the right side of the listener [243].

9.2 Results

SRM was calculated by subtracting the SRT obtained in conditions where maskers were spatially separated from the target (stationary and moving) from the SRT obtained in co-located configurations (target and masker at $0°$ azimuth). A direct comparison between conditions with a stationary masker and moving masker was considered not accurate. Therefore, for the *circular* conditions, the stationary maskers at $20°$, $45°$, and $70°$ azimuth were combined in the so-called *stationary circular masker*. Thus, to analyze SRT, the five stationary Masker Conditions were reduced to three: (1) stationary masker at $0°$, (2) *stationary circular masker*, and (3) *circular moving masker*. To analyze SRM, the three-level variables of masker conditions in the SRT results were further reduced to two levels with masker conditions of (1) *stationary circular masker*, and (2) *circular moving masker*.

For the *radial* conditions, the stationary maskers at 0.8 m, 1.15 m, and 1.5 m were combined in the so-called *stationary radial masker*. The five stationary masker conditions were reduced to three: (1) stationary masker at $0°$, (2) *stationary radial masker*, and (3) *radial moving masker*. To analyze SRM, the three-level variables of masker conditions in the SRT results were further reduced to two levels in SRM with masker conditions of (1) *stationary radial masker*, and (2) *radial moving masker*.

For the *radial* conditions, the "distance law" was applied for both stationary and moving masker conditions, but it was not applied in the *circular* conditions. Therefore, no comparison between *circular* and *radial* conditions was made because noticeable significant differences between masker conditions were expected. Thus, the analysis of *circular* and *radial* conditions was made by separate.

9.2.1 Circular conditions

A repeated-measures analysis of variance (ANOVA) was fitted to the SRT data
with Reverberation (*anechoic* vs. *treated* vs. *untreated*) and Masker Conditions
(*co-located* vs. *stationary* vs. *moving*) as within-subjects variables, while Age
(*young* vs. *elderly*) as the between-subjects variable. The ordinate of Figure 9.6
shows mean SRT data in decibels for both age groups over different masker
conditions, split into reverberant conditions. The left panel contains results from
the *anechoic* condition, the central panel shows results for the *treated* condition
and the right panel shows results for the *untreated* condition. The ANOVA
revealed a significant main effect of Reverberation [$F(2,22) = 596.72$, $p < .001$,
$\eta_p^2 = .72$], reflecting greater SRTs in the *untreated* condition. There was also a
significant main effect of Masker Conditions [$F(2,22) = 91.73$, $p < .001$, $\eta_p^2 =
.34$], in which the co-located condition showed the higher SRTs. The ANOVA
also revealed a significant main effect of Age [$F(1,11) = 72.55$, $p < .001$, $\eta_p^2 =
.53$], where the SRTs were significantly lower in the *young* subjects compared
with the *elderly* subjects (p < .001).
A comparison of data in Figure 9.7 illustrate the significant interactions. There
was an interaction between Reverberation and Masker Conditions [$F(4,44) =
27.32$, $p < .001$, $\eta_p^2 = .29$], reveling the SRTs of all Masker Conditions increase
while more reverberant is the room. There was also an interaction between
Masker Conditions and Age [$F(2,22) = 11.38$; $p < .001$, $\eta_p^2 = .08$], in which
elderly subjects performed greater SRTs for almost all Masker Conditions than
the young subjects. No other significant interactions were found (p > .05).
A comparison of the results for the *co-located* condition with those for spatially separated conditions was analyzed in terms of the amount of SRM. An
ANOVA was fitted to the SRM data with Reverberation (*anechoic* vs. *treated* vs.
untreated) and Masker Conditions (*stationary* vs. *moving*) as within-subjects
variables, while Age (*young* vs. *elderly*) as the between-subjects variable. The
ordinate of Figure 9.8 shows mean SRM data in decibels for both age groups over
different masker conditions, split into reverberant conditions. As expected from
previous research, the ANOVA revealed a significant main effect of Reverberation
[$F(2,22) = 48.48$, $p < .001$, $\eta_p^2 = .57$], reflecting higher SRMs in both the *anechoic*
room over the *treated* room, and in the *treated* room over the *untreated* room.
There was also a significant main effect of Masker Conditions [$F(1,11) = 7.67$, $p
= .018$, $\eta_p^2 = .02$], therefore, a Bonferroni-corrected pairwise comparisons were
applied to examine the difference closely. For the *anechoic* and *treated* conditions
a non-significant difference was found ($p > .05$), but for the untreated condition
a significant difference was found ($p = .048$).

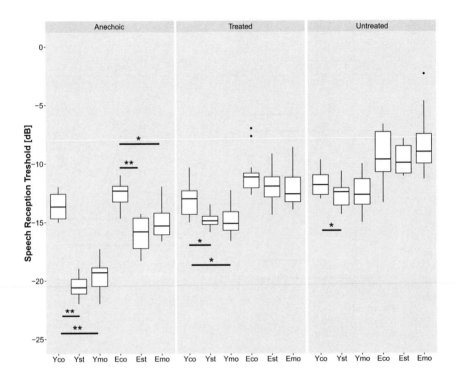

Figure 9.6: Mean SRT in decibels for both age groups over the circular masker conditions, split into the three reverberant conditions. Yco represent SRT of *young* subjects in the co-located condition. Yst is the SRT of *young* subjects in the *stationary circular masker*. Ymo is the SRT of *young* subjects in the *circular moving masker*. Eco represent SRT of elderly subjects in the co-located condition. Est is the SRT of *elderly* subjects in the *stationary circular masker*. Emo is the SRT of *elderly* subjects in the *circular moving masker*. The error bars show 95 % confidence intervals. Asterisks denote significantly different pairs of SRTs with $*$, $p < .05$. and $**$, $p < .001$.

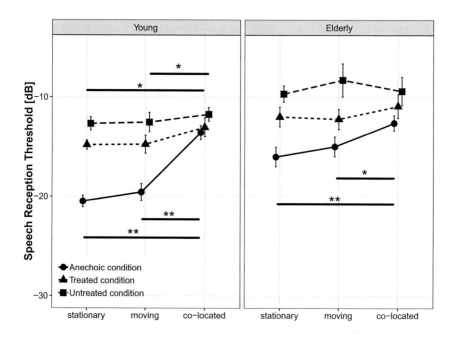

Figure 9.7: SRT data interactions among the three circular masker conditions. The Masker Condition *stationary* represent mean SRT of the *stationary circular masker. moving*, represent the mean SRT of the *circular moving masker. co−located* represent the mean SRT for the masker at the same position than target $(0°)$. The error bars show 95 % confidence intervals. Asterisks denote significantly different pairs of SRMs with ∗, p < .05 and ∗∗, p < .001.

The ANOVA also revealed a significant main effect of Age [$F(1,11) = 15.77$, $p = .002$, $\eta_p^2 = .23$], where the SRMs were significantly higher in the young group compared with the elderly group ($p < .001$).
Additionally, a significant interaction between Reverberation and Age was found [$F(2,22) = 10.83$; $p < .001$, $\eta_p^2 = .14$]. Bonferroni-corrected pairwise comparisons revealed a significant age difference for the *anechoic* and the *untreated* condition ($p < .001$), but a non-significant age effect in the treated condition ($p > .05$). No other significant interactions were found ($p > .05$).

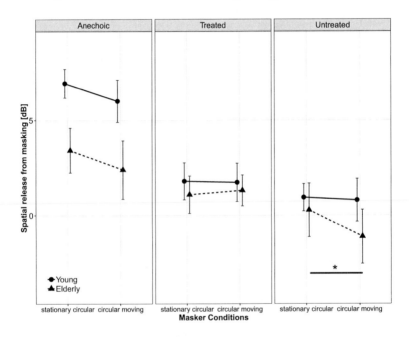

Figure 9.8: Spatial release from masking (SRM) measured for stationary and moving maskers in the *circular* condition. Mean SRMs measured in *anechoic*, *treated* and *untreated* conditions. The error bars show 95 % confidence intervals of the mean. Asterisks denote significantly different pairs of SRMs with ∗, p < .05.

9.2.2 Radial conditions

A repeated-measures ANOVA was conducted to analyze SRM with Reverberation (*anechoic* vs. *treated* vs. *untreated*) and Masker Conditions (*co-located* vs. *stationary* vs. *moving*) as within-subjects variables, while Age (*young* vs. *elderly*) as the between-subjects variable. The ordinate of Figure 9.9 shows mean SRT data in decibels for both age groups over different masker conditions, split into reverberant conditions. The left panel contains results from the *anechoic* condition, the central panel shows results for the *treated* condition and the right panel shows results for the *untreated* condition. The ANOVA revealed a significant main effect of Reverberation [$F(2,22) = 338.17$, $p < .001$, $\eta_p^2 = .76$], reflecting greater SRTs in the *untreated* condition. There was also a significant main effect of Masker Conditions [$F(2,22) = 990.06$, $p < .001$, $\eta_p^2 = .79$], in which co-located was the most effective masker condition. The ANOVA also revealed a significant main effect of Age [$F(1,11) = 85.16$, $p < .001$, $\eta_p^2 = .54$], where the SRTs were significantly lower in the *young* subjects compared with the *elderly* subjects (p < .001).

A comparison of data of Figure 9.10 illustrate the significant interactions. There was an interaction between Reverberation and Masker Conditions [$F(4,44) = 88.47$, $p < .001$, $\eta_p^2 = .40$], reveling the SRTs of all Masker Conditions increase while more reverberant is the room. There was also an interaction between Masker Conditions and Age [$F(2,22) = 12.36$; $p < .001$, $\eta_p^2 = .11$], in which *elderly* subjects performed greater SRTs for almost all Masker Conditions than the *young* subjects. There was also an interaction between Reverberation and Age [$F(2,22) = 3.99$, $p = .03$, $\eta_p^2 = .02$], in which the SRTs of *young* subjects were more spread between Reverberant conditions in comparison with *elderly* subjects. The ANOVA also shown an interaction between Reverberation, Masker Conditions and Age [$F(4,44) = 5.42$, $p < .001$, $\eta_p^2 = .06$].

A comparison of the results for the *co-located* condition with those for spatially separated conditions was analyzed in terms of the amount of SRM. A repeated-measures ANOVA was conducted to analyze SRM with Reverberation (*anechoic* vs. *treated* vs. *untreated*) and Masker Conditions (*stationary* vs. *moving*) as within-subjects variables, while Age (*young* vs. *elderly*) as the between-subjects variable. The ordinate of Figure 9.11 shows mean SRM data in decibels for both age groups over different masker conditions on the abscissa, split into reverberant conditions. The ANOVA revealed a significant main effect of Reverberation [$F(2,22) = 116.06$, p < .001, $\eta_p^2 = .71$], reflecting higher SRMs in both the *anechoic* over the *treated*, and in the *treated* over the *untreated* conditions. There was also a significant main effect of Masker Conditions [$F(1,11) = 5.18$, p = .043, $\eta_p^2 = .01$], therefore, a Bonferroni-corrected pairwise comparisons were applied to examine the difference closely. For the *anechoic* and *untreated* conditions a significant difference was found (p < .001), but for the *treated* condition a non-significant difference was found (p > .05).

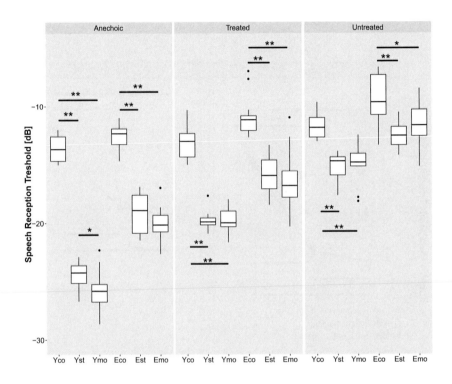

Figure 9.9: Mean SRT in decibels for both age groups over the radial masker
conditions, split into the three reverberant conditions. *Y co* represent
SRT of *young* subjects in the co-located condition. *Y st* is the SRT
of *young* subjects in the *stationary circular masker*. *Y mo* is the
SRT of *young* subjects in the *circular moving masker*. *E co* represent
SRT of elderly subjects in the co-located condition. *E st* is the SRT
of *elderly* subjects in the *stationary circular masker*. *E mo* is the
SRT of *elderly* subjects in the *circular moving masker*. The error
bars show 95 % confidence intervals. Asterisks denote significantly
different pairs of SRTs with ∗, p < .05. and ∗∗, p < .001.

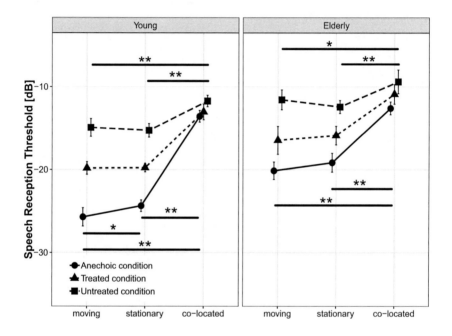

Figure 9.10: SRT data interactions among the three radial masker conditions. The Masker Condition *stationary* represent mean SRT of the *stationary radial masker*. *moving*, represent the mean SRT of the *radial moving masker*. *co−located* represent the mean SRT for the masker at the same position than target (0°). The error bars show 95 % confidence intervals. Asterisks denote significantly different pairs of SRMs with ∗, p < .05 and ∗∗, p < .001.

The ANOVA also revealed a significant main effect of Age [F(1,11) = 18.93, p < .001, η_p^2 = .31], in which the SRMs were significantly higher in the *young* group compared with the *elderly* subjects (p < .001).

For Reverberation and Age, a significant interaction [F(2,22) = 12.05; p < .001] was found. Bonferroni-corrected pairwise comparisons identified a significant age difference for the *anechoic* (p < .001) and the *treated* conditions (p = .007), but a non-significant difference in the *untreated* condition (p > .05). There was also a significant interaction between Reverberation and Masker Conditions [F(2,22) = 17.61; p < .001], reflecting a decrease of SRMs for both masker conditions while more reverberant is the room.

No other significant interactions were found (p > .05).

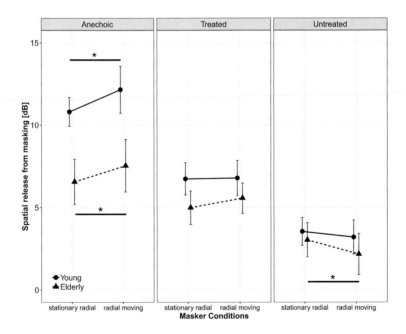

Figure 9.11: Spatial release from masking (SRM) measured for stationary and moving maskers in the *radial* condition. Mean SRMs measured in *anechoic*, *treated* and *untreated* conditions. The error bars show 95 % confidence intervals. Asterisks denote significantly different pairs of SRMs with ∗, p < .05.

9.2.3 IACC results

Figure 9.12 shows the results of the evaluated IACC for all masker positions in all three simulated reverberant conditions. In the *anechoic* condition, highest IACC values all larger than 0.7 were observed. The circular movement showed that the IACC was decreasing for increasing angles, reaching its minimum at 76°. Data points of the *circular moving masker* also indicated the limited angular resolution of the chosen HRTF dataset (3° x 3° in azimuth and zenith angles), resulting in triplets of identical IACC values. For the *radial moving masker*, all evaluated data points lead to identical IACC values as the corresponding BRIRs of the *anechoic* condition only differed in amplitude. The effect of the air absorption, which was considered in the simulation, has a negligible effect on the BRIRs and their corresponding IACC values. The values for the two reverberant conditions including room reflections show that the IACC is lower for all positions. It is noteworthy that in case of the circular maskers, IACC values were rather high for the frontal position (above 0.6), but substantially decreased for larger angles (below 0.3 for 70°). When comparing both reverberant conditions, the *untreated*, which has a lower direct-to-reverberant ratio than the *treated*, also has lower IACC values. In this case, the effect of a decreasing direct-to-reverberant ratio on the IACC due to increasing the sound source to receiver distance, was also visible for the radial conditions.

9.3 Discussion and Conclusions

The current experiment examined situations in which the listener must perform a speech-in-noise task in the presence of a masker located at different positions or moving. This experiment focused on the benefit of spatial separation. As is known, the SRM performance in reverberant conditions is different for *young* and *elderly* subjects [164, 225], but how a moving masker affect the SRM under different reverberant conditions has not been investigated yet.

The main objective of the present experiment was to establish the role played by the movement of the masker in realistic listening situations and to determine if exist differences between *young* and *elderly* subjects in such circumstances.

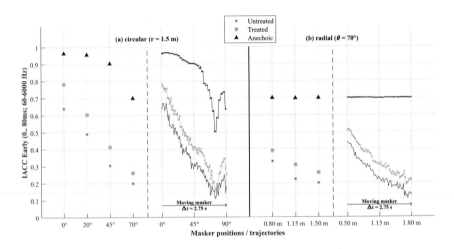

Figure 9.12: Interaural cross-correlation coefficient evaluated for the binaural room impulse responses corresponding to the test conditions. The left part shows the results for the four sampled masker positions and for continuous masker movement both on the circular trajectory from 0° to 90° azimuth. The right part shows the results for the three sampled masker positions and for continuous radial masker movement at 70° azimuth distances from 0.5 m to 1.8 m.

9.3.1 Effect of moving masker in circular conditions

In the *anechoic* condition it is possible to observe only a numerical tendency, as the SRM of *circular moving masker* is slightly lower than the *stationary circular masker* for both age groups, however, the difference is not significant.

In the *treated* condition, no significant differences between *stationary circular* and *circular moving* maskers were found in both age groups. The slight SRM benefits observed for the*stationary circular masker* in the *anechoic* condition were annulated by the room reflections.

In the *untreated* condition, a significant difference between *stationary circular* and *circular moving* maskers was found, but only in *elderly* subjects. The circular movement of the masker, combined with a high level of reverberation, can be considered as a more complex listening situation for the *elderly* subjects than stationary masker conditions. The situation seems to be so harmful that the mean SRMs reached a negative value, which means that the mean SRT in the *circular moving masker* was even lower than the mean SRT in the *co-located* condition, where there is no spatial separation advantage. The effect due to the movement of the *circular moving masker* was around 1.5 dB of difference with respect to the *stationary circular masker*. The specific factor of the movement that produced the reduction on SRM is unclear.

From the IACC evaluation, it can be observed that the interaural coherence decreases for all positions in the reverberant conditions. It is known that the general reduction of interaural coherence reduces the speech intelligibility [150]. Nevertheless, it needs to be researched more thoroughly in how the *circular moving masker* is more distracting than the *stationary circular masker* when the IACC of the target speaker is also very low. When comparing all three reverberant conditions it can be concluded that a high interaural coherence is more beneficial for younger listeners, nevertheless, their advantage is lost as soon as the binaural coherence decreases in reverberant environments. Regarding the comparison between *stationary circular* and *circular moving* maskers, the IACC analysis, however, does not allow any further explanations.

9.3.2 Effect of moving masker in radial conditions

In the *anechoic* condition, a significant moving effect was found. For both age groups, the *stationary radial masker* has a lower SRM than the *radial moving masker*. The favorable effect due to the motion of the masker was around 1.5 and 1 dB for *young* and *elderly* subjects, respectively. In case of the *radial moving masker*, the sound pressure decreases proportionally to the inverse of the distance between sound source and receiver, while for the given situation, the IACC of the target source and the masker is identical for all situations, which suggests an improvement in SRM due to the radial movement of the masker. In other words, the radial movement of the masker could generate larger differences in SNR at the better ear, resulting in higher benefits in SRM.

In the *treated* condition, no significant difference between *stationary radial* and *radial moving* masker was found. The addition of room reflections led to a substantial reduction of the SRM especially for *young* subjects. The reduction of the IACC of the target as well as for the masker in comparison with the *anechoic* environment supports previous findings that interaural coherence is important to benefit from spatial separation [150, 156, 209]. Therefore, the SRM benefits are independent of the radial movement of the masker in such reverberant condition.

In the *untreated* condition, a significant difference between *stationary radial* and *radial moving* masker was found, but only for the *elderly* subjects with (1) *stationary circular masker*, $M = 3.03$ dB, $SD = 1.0$ dB, and (2) *circular moving masker*, $M = 2.17$ dB, $SD = 1.19$ dB. Here, it can also be assumed that the radial movement of the masker in a high reverberant environment was a more complex listening situation for the *elderly* subjects than the *stationary radial* masker. The effect due to the radial movement of the masker was around 1.0 dB of difference with respect to the *stationary radial masker*.

In contrast to the *anechoic* condition (direct sound), the reduction of IACC is more relevant than the reduction of the masker's sound pressure in case of the *radial moving masker*, leading to a decrease of the SRM for *elderly* subjects. Additionally, in both reverberant conditions, the sound pressure of the masker does not decay according to the distance law, the additional energy in comparison with the *anechoic* condition is mostly uncorrelated energy, thus reducing the potential for spatial separation. To define a model for this correlation, a more thorough investigation is required at this point, including an assessment of the temporal variation of the SNR and other potentially relevant parameters, such as loudness or the binaural level.

9.3.3 Effect of age in circular conditions

In the *anechoic* condition, a significant Age effect was found, where *elderly* subjects reported lower SRM than *young* subjects. These results are similar to those of Dubno et al. (2002) [77] that investigated SRM in younger listeners with normal hearing, older adults with normal hearing and older listeners with

hearing loss. They found a significant difference between subjects in both the
younger listeners over older listeners with normal hearing, and older listeners
with normal hearing over older listeners with hearing loss. Conversely, Dubno
et al. (2008) [75], that also investigated SRM in young and elderly subjects
with similar test design as the previous study, found only a trend of SRM where
young subjects had higher benefits than elderly listeners, but the difference
was not significant. However, in contrast to the current experiment, in Dubno
et al. (2008) [75] the speech and the masker were low-pass filtered at 3 kHz
to minimize differences in audibility of higher frequency cues between the two
age groups. Therefore, the significant difference between subjects found in the
anechoic condition could be due to the reduced ability of *elderly* subjects in
using the interaural difference cues [75], especially for the ILD (head shadow)
which is directly related to the frequency range where the *elderly* subjects have
hearing-impairments. In this experiment, a difference of about 3.5 dB due to this
detrimental effect was observed.

In the *treated* condition there were no significant differences between *young*
and *elderly* subjects. It is known that poor room acoustics affects negatively
the SRM [124]; therefore, in comparison with the *anechoic* condition, a lower
SRM in the *treated* condition was expected. Nevertheless, it was not clear if the
reduction in SRM would be similar for both age groups. Both *young* and *elderly*
subjects had a reduction of SRM in the *treated* condition in comparison with the
anechoic condition, but the detrimental effect was deeper for the *young* subjects
($\sim$5 dB for the *stationary circular masker* and $\sim$4 dB for the *circular moving
masker*) than for the *elderly* subjects ($\sim$2 dB for the *stationary circular masker*
and $\sim$1 dB for the *circular moving masker*). Marrone et al. [164] investigated
SRM among young and elderly subjects under both low and high reverberant
conditions. They found a significant difference between age groups in the low
reverberant condition, however, the maskers were also a speech signal similar to
the target, and its locations were to either side ($\pm 90°$) to diminish the use of
the head shadow effect. In the current experiment, both age subjects performed
similar SRMs in the *treated* condition, thus, is possible to assume that this low
reverberant condition does not enlarge strongly the detrimental effect that has
the hearing-impairment of *elderly* subjects in reducing their ability to use ILDs.
Conversely, the *young* subjects were largely affected by the reverberant condition
with around 4-5 dB of SRM reduction in both *stationary circular* and *circular
moving* maskers, thus obtaining similar results than *elderly* subjects. Therefore,
a more thorough investigation is required to know if reverberation has a similar
detrimental effect as the hearing-impairments of the *elderly* subjects over the
SRM in *circular* conditions.

In the *untreated* condition, an unexpected significant age difference in the *circular moving* masker was found, where the *elderly* subjects performed lower SRMs with (1) *stationary circular masker*, $M = 0.27$ dB, $SD = 1.34$ dB, and (2) *circular moving masker*, $M = $ -1.12 dB, $SD = 1.36$ dB. For the *circular moving masker*, the mean SRM of the *young* subjects decreases from the *treated* condition to the *untreated* condition by 1 dB, meanwhile the mean SRM of the *elderly* subjects decrease from the *treated* condition to the *untreated* condition by 2.5 dB. This allows the conclusion that, for *elderly* subjects, the acoustic scene becomes more complex and demanding when the masker is moving and the reverberation increase.

In general, significant differences were found between *young* and *elderly* subjects in several spatial separation conditions tested in this experiment. Even so, it should be clarified that age effects were not free from influences of hearing-impairment, thus, the interpretation of the results cannot be attributed to age only [164]. The age effect could be due to declines in cognition functions (such as working memory), attentional control, processing speed, or declines in auditory functions [176]. By excluding the influences of hearing–impairment to focus only on the age effect, could not represent faithfully one of the biggest group with hearing difficulties without being hearing-aid wearers. Nevertheless, due to the design of this experiment, is possible to assume that differences could be related to auditory processing alterations since the task was not related with cognition processing and all *elderly* subjects were tested to be normal in the *DemTect* test.

9.3.4 Effect of age in radial conditions

In the *anechoic* condition, a significant age effect was found for *young* subjects with (1) *stationary radial masker*, $M = 10.81$ dB, $SD = 0.84$ dB, and (2) *radial moving masker*, $M = 12.15$ dB, $SD = 1.36$ dB, and for *elderly* subjects with (1) *stationary radial masker*, $M = 6.56$ dB, $SD = 1.32$ dB, and (2) *radial moving masker*, $M = 7.54$ dB, $SD = 1.52$ dB. The *elderly* subjects have a lower SRM than the *young* subjects, with a difference of around 4.5 dB for both masker conditions. The difference between subjects could be due to the reduction that the *elderly* subjects have in using the monaural and interaural difference cues, especially the differences in SNR at the better ear.

For the *treated* condition, it was expected a reduction in the SRM for both age groups, but again it was not clear if the reduction in SRM would be similar for both age groups under the *radial* conditions. Unlike the *circular* conditions, there was a reduction in SRM, but the significant difference between age groups remained. It was previously suggested that there is an improvement of the SRM benefit due to the radial movement of the masker, thus, even in this reverberant condition, *young* subjects could have benefited to a greater extent of the shadow effect than *elderly* subjects.

In the *untreated* condition, there were no significant differences between *young* and *elderly* subjects. Therefore, is possible to assume that this high reverberant condition does not enlarge the detrimental effect that has the hearing-impairment of *elderly* subjects in reducing their ability to use ILDs. Consequently, a more thorough investigation is required to know if reverberation has a similar detrimental effect as the hearing-impairments of the *elderly* subjects over the SRM in *radial* conditions.

9.3.5 Conclusion

The main purpose of the present experiment was to establish the effect of a moving masker, either *circular* or *radial*, in an SRM task between *young* and *elderly* subjects across different reverberant conditions.

For *circular* conditions, a slight tendency in the *anechoic* condition was found, with the SRM of *circular moving masker* being lower than the SRM of the *stationary circular masker*. Conversely, and as was expected, in the *untreated* condition the *elderly* subjects got a lower SRM in the *circular moving masker* than the *stationary circular masker*, which suggests that the circular movement of the masker is a more challenging listening situation than stationary conditions.

For *radial* conditions, the SRM of the *radial moving masker* in the *anechoic* condition was higher than the *stationary radial masker*, showing that the radial movement of the masker produces larger differences in SNR at the better ear. In the *untreated* condition, a significant difference in *elderly* subjects was found, where the SRM of the *radial moving masker* was lower than the *stationary radial masker*. This shows that, for the *elderly* subjects, even in the situation of an increasing SNR during the *radial moving masker* condition (which expects higher SRM), the fact that the situation was dynamic also lead to a more challenging listening situation than stationary conditions.

Consistent with previous research, *young* subjects demonstrated higher SRM than *elderly* subjects, especially for conditions with a high level of interaural coherence of target and masker, for example in the *anechoic* condition. The effect due to age, however, was inconclusive. Typical auditory processing alterations of the *elderly* subjects were considered as the possible explanation for the findings; however, different experimental designs are required to answer this question with certainty.

In general, the results of the presented experiment show that *elderly* subjects have more difficulties to handle moving maskers than stationary maskers in poor reverberant conditions.

10

Conclusion and Outlook

It is known that listeners with hearing impairments have difficulties understanding speech in the presence of background noise. Despite the high technology presented by the new hearing aids, listeners with hearing impairments still complain about their capacity to understand speech under complex noise environments. The hypothesis of this thesis was that the current listening tests, performed to assess speech-in-noise perception, are not representative of real-life situations. Most research has concentrated on the listener's ability to understand speech in the presence of maskers such as noise or speech, but they have largely focused only on stationary sound sources. Nevertheless, as we know, in real-life listening situations we are confronted with multiple stationary and moving sound sources that disturb our speech perception.

This thesis provides insights into the listener's speech perception capacity in cases with moving maskers. Furthermore, comparing moving and stationary maskers, this thesis provides information on whether the hearing system processes these two situations differently. In other words, the objective was to know whether the evaluation of a moving masker reports differents speech reception thresholds (SRTs) and spatial release from masking (SRM) than stationary maskers.

In chapter 6 a virtual acoustic environment was simulated to assess SRM of moving maskers in an anechoic condition. Maskers moving away and toward the target position were compared, but nevertheless, no differences between the two trajectories were found. Besides, different degrees of movements were evaluated between 15° and 90°, where a clear pattern was reflected with masker moving 15° achieving the lowest SRM and masker moving 90° the highest SRM. However, it is known that many different factors can affect the speech-in-noise perception such as the measurement paradigm, head movements, room acoustic conditions, the use of individual HRTFs, the type of the masker and its spatial distribution, among others [34]. Therefore, in this thesis, several factors to assess speech-in-noise perception in different moving masker conditions were taken into account.

Due to the relevance of the SRM on speech perception, several models have been created to predict SRM for diverse spatial positions of the masker and for different types of maskers. However, so far, none of the models has taken into account maskers in movement. In chapter 7, a mathematical model to predict SRM for maskers with movements in azimuth (only from $-90°$ to $+90°$) was presented. It was possible to observe different SRM between stationary and moving maskers, especially for short movements.

While head movement has been shown to improve sound localization accuracy [230], how it affects performance in SRM tasks remains unknown. In chapter 8, an experiment comparing static and dynamic reproduction methods was presented. The experiment investigated whether listeners used their head movements to maximize their intelligibility during a listening test with a moving masker. Differences only in the stationary masker cases were observed, showing higher SRTs (worst intelligibility) in cases with head movements (dynamic reproduction) than cases without head movements (static reproduction). In cases with a moving masker, no differences were found; therefore, listeners could have used their head movements more effectively to resolve the rapidly changing and ambiguous binaural cues while the masker was moving. Besides, between masker moving away and toward the target position, no differences were found when the dynamic reproduction was implemented.

The comparison of stationary and moving maskers with the use of dynamic reproduction was found to be significantly different. Nevertheless, it is known that the use of mismatched HRTFs data-sets, such as HRTFs measured from an artificial head, can result in unwanted artifacts in coloration and localization. For that reason, the use of individual HRTFs and an artificial head HRTF was evaluated. Results showed significant differences between individual HRTF and artificial head HRTF, reflecting in general higher SRTs for artificial head HRTF than individual HRTF, but no differences in SRM. Besides, no differences in the use of head movements were found. In addition, the comparison between stationary and moving masker showed a significant difference only for static reproduction and individual HRTF.

Another relevant factor that affects SRM is room acoustics. For that reason, in chapter 9, an analysis comparing stationary maskers with two different moving masker (circular and radial), among different reverberant conditions (anechoic, treated and untreated) was presented. This analysis was carried out within two groups of subjects: young and elderly subjects (no hearing aid users). For both circular and radial conditions in the untreated room, a significant difference between stationary and moving masker was found. This difference was observed in the elderly subjects, who are one of the main groups in which this thesis is oriented. These findings meaning that, in high reverberant conditions, elderly subjects have more difficulties to understand speech with a moving masker than stationary maskers.

To summarize, moving versus stationary maskers under different conditions were assessed in all the experiments evaluated in this thesis. Significant differences in several cases and conditions were found, revealing that the auditory system assesses differently the moving maskers than the stationary maskers. Therefore, the inclusion of moving conditions in clinical listening tests is recommended, in order to assess speech-in-noise perception in a more realistic environment. Nevertheless, the role that visual cues play on head movements during a listening test with moving maskers needs to be researched more thoroughly. Furthermore, a more complete model of SRM to predict moving maskers with every movement in 360° under different reverberant conditions could bring more insight into the dynamic binaural mechanism. In addition, evaluating hearing aids wearers under moving sound sources conditions may provide information related to a better fitting of hearing aids.

List of Acronyms

ANOVA Analysis of Variance

BRIR Binaural Room Impulse Response

CI Cochlear Implant

CRM Coordinated Response Measure

DS Direct Sound

DSRM Dynamic Spatial Release From Masking

ER Early Reflections

FST Freiburg Speech Test

GUI Graphical User Interface

HINT Hearing in Noise Test

HL Hearing Level

HpTF Headphone Transfer Function

HRTF Head-Related Transfer Function

IACC Interaural Cross-Correlation

ILD Interaural Level Difference

IPD Interaural Phase Differences

ITA Institute of Technical Acoustics

ITD Interaural Time Difference

LISN Listening in Spatialized Noise

LR Late Reverberation

LTASS Long-Term Average Speech Spectrum

MAMA Minimum Audible Movement Angle

OLSA Oldenburg Sentence Test

PTA Pure-Tone Audiometry

REA Right-Ear Advantage

R-HINT-E Realistic Hearing in Noise Test Environment

RIR Room Impulse Response

RT Reverberation Time

SIN Speech in Noise

SNR Signal to Noise Ratio

SPIN Speech Perception in Noise

SPL Sound Pressure Level

SPRINT Speech Recognition in Noise Test

SRM Spatial Release from Masking

SRT Speech Reception Threshold

SUN Speech Understanding in Noise

VA Virtual Acoustics

VAE Virtual Acoustic Environment

WIN Words-in-Noise

WRS Word Recognition Score

List of Figures

2.1   Auditory Sensation Area. Adopted from Zwicker and Fastl [267].   9

2.2   Anatomy of the pinna. 10

2.3   Head-related coordinate system. Adopted from [81]. 11

2.4   Descriptive definition of the free-field HRTF. Adopted from [81].   12

2.5   Descriptive scheme of three sound source positions with different ITDs. Adopted from [81]. 13

2.6   Descriptive scheme of the frequency dependence of ILD with low frequencies. Adopted from [81]. 14

2.7   Descriptive scheme of the frequency dependence of ILD with high frequencies. Adopted from [81]. 14

2.8   Descriptive scheme of five sound sources positions in the cone of confusion with identical ITD and ILD. Adopted from [81]. 15

2.9   Sound source positions where ITD and ILD are zero, provoking confusion on the real localization of the source. Adopted from [81]. 15

2.10   Schematic representation of the "Head Shadow" effect. L represents the left ear and R the right ear. The level ratio between the target signal regardless of the masker signal in the left ear (SNR L) is larger than for the right ear (SNR R). For that reason, in this example, the left ear has an advantage in speech-in-noise perception. 19

2.11   Measurement setup of the arc and an artificial head on a rotating turn table at ITA, RWTH Aachen University. 24

3.1   Psychometric functions of word recognition performance measured in percent correct (y-axis) for a listener as a function of presentation level (x-axis). The dashed line indicates the 50 % point. The SRT represent the level (SNR) at which the listener is able to recognize the 50 % of the words. 27

3.2   Changes in SNR during an adaptive procedure to track the SRT (SNR at 50% speech intelligibility). 29

3.3   Example of masker configuration to calculate SRM. 33

4.1   ITA sound-attenuated hearing booth 41

4.2   High quality headphones Sennheiser HD 650   41

4.3   Camera OptiTrack Flex 13, with 1.3 million pixels of resolution
      and 120 FPS sample rate   43

4.4   Headphones Sennheiser HD 650 with the rigid body on the top .   44

5.1   Noise power distribution per frequencies between 20 Hz and 20 kHz. 48

5.2   Sequence of the digit-triplet test. (a) Illustrations of the digit-
      triplet stimulus playback stream. (b) Mean playback time of all
      digits in each position of the triplet. (c) Timeline of the noise
      masker.   51

6.1   Graphical representation of all cases in this study and the masker
      position/movement. Five different masker movements (15°, 30°,
      45°, 60° and 90°) for two trajectories (away and toward the target
      position) were evaluated. T denotes the target sound source
      location and M is masker position/movement during the trial. . .   55

6.2   Speech reception threshold (SRT) measured for eleven different
      masker conditions: stationary at 0°, and moving away and toward
      the target position (15°, 30°, 45°, 60° and 90°).   56

6.3   Dynamic spatial release from masking (DSRM) measured for five
      different masker movements (15°, 30°, 45°, 60° and 90°) and for
      two trajectories: away and toward to the target position.   57

7.1   Target/Masker (T/M) configurations to calculate SRM. Masker
      sound sources were either located at stationary positions or moving
      along a trajectory.   62

7.2   Graphical representation of all conditions tested in the current
      experiment showing the masker's location/trajectory. T represents
      the target sound source and M is the masker source position or
      movement during the trial. * represent both maskers stationary
      at +90°   64

7.3   SRM prediction for both *stationary models* when target is at 0°
      and the masker at different locations between 0° and 90° together
      with the accumulated SRM of both *stationary models* when the
      masker move from 0° to 90°.   65

7.4   DSRM is shown for moving maskers to 15, 30, 45, 60 and 90
      degrees, together with predicted SRM of both *stationary models*
      for the $+\theta/+\theta$ masker configuration.   67

7.5 DSRM binaural contribution for maskers moving 15, 30, 45, 60 and 90 degrees, together with predicted SRM of both *stationary models*. The Binaural component is examined with the symmetric masker configuration $-\theta/+\theta$. 69

7.6 DSRM better-ear contribution for maskers moving 15, 30, 45, 60 and 90 degrees, together with predicted SRM of both *stationary models*. The better-ear component is examined with the masker configuration described in equation 7.3. 70

8.1 ITD as a function of azimuth angle and time on the condition masker moving 90°, with a 3° resolution. The ITD was calculated by cross-correlating the HRTFs after applying a bandpass filter with cutoff frequencies at 200 Hz and 1000 Hz, specifically for the useful frequency range from the speaker whose fundamental frequency was around 200 Hz. 75

8.2 ILD as a function of azimuth angle and time on the condition masker moving 90°, plotted for 2000, 3000, 4000, and 5000 Hz. As the masker moves from 0° to 90°, ILD increases and reaches peak at different angles across the frequencies plotted. 76

8.3 Stationary and moving masker conditions. Left side shows the spatial position of the masker for the five stationary cases. The right side shows cases with masker moving away from 0° to 90° and moving toward from 90° to 0°. 77

8.4 Predicted SRM in the five stationary spatial configurations: $T_0 M_0$, $T_0 M_{20}$, $T_0 M_{45}$, $T_0 M_{70}$, and $T_0 M_{90}$. 78

8.5 Psychometric functions of the three stationary masker conditions (20°, 45° and 70°) together with the predicted psychometric function of a masker moving 90° (away or toward the target position). 80

8.6 Psychometric functions of both moving away and toward masker conditions together with the predicted psychometric function of a masker moving 90°. 81

8.7 Speech reception thresholds at 50 % speech intelligibility measured for each masker condition: (a) stationary at 0°, (b) stationary at 20°, (c) stationary at 45°, (d) stationary at 70°, (e) moving away, (f) moving toward, and (g) stationary at 90°. In each masker condition, the SRT was plotted separately for with and without listener head movement tracked. Asterisks denote the significantly different pairs of SRTs measured with versus without head movement at $p < .05$. The error bars show 95 % confidence intervals. Asterisks denote the significantly different pairs of SRTs with *, $p < .05$ and **, $p < .001$. 82

8.8 SRM measured for masker conditions: (1) stationary at 20°, (2) stationary at 45°, (3) stationary at 70°, (4) moving away, (5) moving toward, and (6) stationary at 90°. Mean SRMs measured in virtual acoustic environments with static binaural reproduction (listener head movement was not used to update head-related transfer functions, HRTFs) and with dynamic binaural reproduction (listener head movement was used to update HRTFs in real-time). The error bars show 95 % confidence intervals. . . . 84

8.9 Slope measured for the predicted condition vs. slope of the masker moving away vs. slope of the masker moving toward. Mean slope measured in virtual acoustic environments with static and dynamic binaural reproduction. The error bars show 95 % confidence intervals. Asterisks denote the significantly different pairs of slopes with *, $p < .05$. 85

8.10 SRT measured for the predicted condition vs. SRT of the masker moving away vs. SRT of the masker moving toward. Mean SRT measured in virtual acoustic environments with static and dynamic binaural reproduction. The error bars show 95 % confidence intervals. 85

8.11 Head movements on the horizontal plane of the case masker stationary at 90°. (a) Show movements to the right and its corresponding SNRs during the adaptive procedure to track the SNR at 50 % speech intelligibility, (b) show movements to the left and its corresponding SNRs, and (c) show bi-directional movements and its corresponding SNRs. The black line represents the lowest SRT reached in each case and their correspondent head movements, meanwhile, the gray line represents the rest of listeners' head movements and its corresponding SNRs during the adaptive procedure. 89

8.12 Head movements on the horizontal plane of the case moving masker away from the target. (a) Show movements to the left and its corresponding SNRs during the adaptive procedure to track the SNR at 50 % speech intelligibility, and (b) show bi-directional movements and its corresponding SNRs. The black line represents the lowest SRT reached in each case and their correspondent head movements, meanwhile, the gray line represents the rest of listeners' head movements and its corresponding SNRs during the adaptive procedure. 90

8.13 Head movements on the horizontal plane of the case moving masker toward to the target. (a) Show movements to the right and its corresponding SNRs during the adaptive procedure to track the SNR at 50 % speech intelligibility, (b) show movements to the left and its corresponding SNRs, and (c) shows bi-directional movements and its corresponding SNRs. The black line represents the lowest SRT reached in each case and their correspondent head movements, meanwhile, the gray line represents the rest of listeners' head movements and its corresponding SNRs during the adaptive procedure. 92

8.14 Spatial location of masker assessed in the experiment. The M represents the position or movement of the masker. 94

8.15 Speech reception thresholds at 50 % speech intelligibility measured for each masker condition: (a) stationary at 0°, (b) moving 90°, and (c) stationary at 90°. In each masker condition, the SRT was plotted separately for both HRTF used and for both binaural reproduction methods. S_AH represent static reproduction and the used of artificial head HRTF. S_IND represent static reproduction and the used of individual HRTF. D_AH represent dynamic reproduction and the used of artificial head HRTF. D_IND represent dynamic reproduction and the used of individual HRTF. The error bars show 95 % confidence intervals. Asterisks denote the significantly different pairs of SRTs with *, $p < .05$. 96

8.16 Spatial release from masking (SRM) measured for masker moving 90°, and stationary at 90°. Mean SRMs measured in virtual acoustic environments with static and dynamic binaural reproduction, and for both HRTFs used in the binaural synthesis (artificial head HRTF and individual HRTF). The error bars show 95 % confidence intervals. 97

9.1 Spatial configuration of the listener and talker in the virtual room. 105

9.2   Simulated reverberation times T_{20} for the room conditions *treated* and *untreated*. Values were averaged for 12 evaluated room impulse responses including six receiver positions and two sound source positions.   105

9.3   Stationary and moving maskers in the *circular* condition. The left side shows the four angular positions of the masker on the stationary conditions. The right side shows the condition with *circular moving masker* (from 0° to 90° azimuth).   107

9.4   Illustrations of time events during playback of both target and masker streams. In the virtual scene, masker movements start after the leading tone. (a) Illustrations of the digit-triplet stimulus playback stream. (b) Mean angular positions of the continuous circular masker movement, indicating time instances of individual digit playback. (c) Mean radial positions of the continuous radial masker movement indicating time instances of individual digit playback. (d) Timeline for stimulus streams.   108

9.5   Stationary and moving maskers in the *radial* condition. The left side shows the three radial positions of the masker on the stationary conditions. The right side shows the condition with *radial moving masker* (from 0.5 m to 1.8 m at 70° azimuth). . .   108

9.6   Mean SRT in decibels for both age groups over the circular masker conditions, split into the three reverberant conditions. *Yco* represent SRT of *young* subjects in the co-located condition. *Yst* is the SRT of *young* subjects in the *stationary circular masker. Ymo* is the SRT of *young* subjects in the *circular moving masker. Eco* represent SRT of elderly subjects in the co-located condition. *Est* is the SRT of *elderly* subjects in the *stationary circular masker. Emo* is the SRT of *elderly* subjects in the *circular moving masker*. The error bars show 95 % confidence intervals. Asterisks denote significantly different pairs of SRTs with ∗, p < .05. and ∗∗, p < .001.   113

9.7   SRT data interactions among the three circular masker conditions. The Masker Condition *stationary* represent mean SRT of the *stationary circular masker. moving*, represent the mean SRT of the *circular moving masker. co − located* represent the mean SRT for the masker at the same position than target (0°). The error bars show 95 % confidence intervals. Asterisks denote significantly different pairs of SRMs with ∗, p < .05 and ∗∗, p < .001.   114

9.8   Spatial release from masking (SRM) measured for stationary and moving maskers in the *circular* condition. Mean SRMs measured in *anechoic, treated* and *untreated* conditions. The error bars show 95 % confidence intervals of the mean. Asterisks denote significantly different pairs of SRMs with $*$, p $<$ .05.   115

9.9   Mean SRT in decibels for both age groups over the radial masker conditions, split into the three reverberant conditions. *Yco* represent SRT of *young* subjects in the co-located condition. *Yst* is the SRT of *young* subjects in the *stationary circular masker*. *Ymo* is the SRT of *young* subjects in the *circular moving masker*. *Eco* represent SRT of elderly subjects in the co-located condition. *Est* is the SRT of *elderly* subjects in the *stationary circular masker*. *Emo* is the SRT of *elderly* subjects in the *circular moving masker*. The error bars show 95 % confidence intervals. Asterisks denote significantly different pairs of SRTs with $*$, p $<$ .05. and $**$, p $<$ .001.   117

9.10   SRT data interactions among the three radial masker conditions. The Masker Condition *stationary* represent mean SRT of the *stationary radial masker*. *moving*, represent the mean SRT of the *radial moving masker*. *co − located* represent the mean SRT for the masker at the same position than target ($0°$). The error bars show 95 % confidence intervals. Asterisks denote significantly different pairs of SRMs with $*$, p $<$ .05 and $**$, p $<$ .001.   118

9.11   Spatial release from masking (SRM) measured for stationary and moving maskers in the *radial* condition. Mean SRMs measured in *anechoic, treated* and *untreated* conditions. The error bars show 95 % confidence intervals. Asterisks denote significantly different pairs of SRMs with $*$, p $<$ .05.   119

9.12   Interaural cross-correlation coefficient evaluated for the binaural room impulse responses corresponding to the test conditions. The left part shows the results for the four sampled masker positions and for continuous masker movement both on the circular trajectory from $0°$ to $90°$ azimuth. The right part shows the results for the three sampled masker positions and for continuous radial masker movement at $70°$ azimuth distances from 0.5 m to 1.8 m.   121

List of Tables

5.1 Parameters of psychometric functions in the digit-triplet test created by [265] and from the present thesis, both digit- and triplet-wise scoring. The mean SRT of 50% intelligibility and the mean slope of the psychometric functions are listed with standard deviation σ. 50

9.1 Stationary and moving masker definitions for *circular* and *radial* conditions. 109

Bibliography

[1] ISO 3382-1. *Acoustics—Measurement of room acoustic parameters—Part 1: Performance spaces.* 2009.

[2] Michael A Akeroyd. "The psychoacoustics of binaural hearing: La psicoacústica de la audición binaural." In: *International journal of audiology* 45.sup1 (2006), pp. 25–33.

[3] Michael A Akeroyd et al. "International Collegium of Rehabilitative Audiology (ICRA) recommendations for the construction of multilingual speech tests: ICRA Working Group on Multilingual Speech Tests." In: *International journal of audiology* 54.sup2 (2015), pp. 17–22.

[4] Claude Alain. "Breaking the wave: effects of attention and learning on concurrent sound perception." In: *Hearing research* 229.1-2 (2007), pp. 225–236.

[5] Jont B Allen and David A Berkley. "Image method for efficiently simulating small-room acoustics." In: *The Journal of the Acoustical Society of America* 65.4 (1979), pp. 943–950.

[6] Kachina Allen, David Alais, and Simon Carlile. "Speech intelligibility reduces over distance from an attended location: evidence for an auditory spatial gradient of attention." In: *Perception & Psychophysics* 71.1 (2009), pp. 164–173.

[7] Kachina Allen, Simon Carlile, and David Alais. "Contributions of talker characteristics and spatial location to auditory streaming." In: *The Journal of the Acoustical Society of America* 123.3 (2008), pp. 1562–1570.

[8] S3 ANSI. "American National Standard Methods for Calculation of the Speech Intelligibility Index." In: *Norme ANSI* (1997).

[9] Tanya L Arbogast, Christine R Mason, and Gerald Kidd Jr. "The effect of spatial separation on informational masking of speech in normal-hearing and hearing-impaired listeners." In: *The Journal of the Acoustical Society of America* 117.4 (2005), pp. 2169–2180.

[10] Lukas Aspöck et al. "A real-time auralization plugin for architectural design and education." In: *10.14279/depositonce-4103* (2014).

[11] John Charlton Baird and Elliot Jason Noma. *Fundamentals of scaling and psychophysics*. John Wiley & Sons, 1978.

[12] Enrique Salesa Batlle, Enrique Perelló Scherdel, and Alfredo Bonavida Estupiñá. *Tratado de audiologia*. Elsevier Health Sciences, 2013.

[13] Randall C Beattie. "Word recognition functions for the CID W-22 test in multitalker noise for normally hearing and hearing-impaired subjects." In: *Journal of Speech and Hearing Disorders* 54.1 (1989), pp. 20–32.

[14] Durand R Begault, Elizabeth M Wenzel, and Mark R Anderson. "Direct comparison of the impact of head tracking, reverberation, and individualized head-related transfer functions on the spatial perception of a virtual speech source." In: *Journal of the Audio Engineering Society* 49.10 (2001), pp. 904–916.

[15] Georg v Békésy. "A new audiometer." In: *Acta Oto-Laryngologica* 35.5-6 (1947), pp. 411–422.

[16] John Bench, Åse Kowal, and John Bamford. "The BKB (Bamford-Kowal-Bench) sentence lists for partially-hearing children." In: *British journal of audiology* 13.3 (1979), pp. 108–112.

[17] Charles I Berlin et al. "Dichotic Right Ear Advantage in Children 5 to 131." In: *Cortex* 9.4 (1973), pp. 394–402.

[18] Marco Berzborn et al. "The ITA-Toolbox: An open source MATLAB toolbox for acoustic measurements and signal processing." In: *Proceedings of the 43th Annual German Congress on Acoustics, Kiel, Germany*. 2017, pp. 6–9.

[19] Virginia Best et al. "The influence of non-spatial factors on measures of spatial release from masking." In: *The Journal of the Acoustical Society of America* 131.4 (2012), pp. 3103–3110.

[20] Rainer Beutelmann and Thomas Brand. "Prediction of speech intelligibility in spatial noise and reverberation for normal-hearing and hearing-impaired listeners." In: *The Journal of the Acoustical Society of America* 120.1 (2006), pp. 331–342.

[21] Rainer Beutelmann, Thomas Brand, and Birger Kollmeier. "Revision, extension, and evaluation of a binaural speech intelligibility model." In: *The Journal of the Acoustical Society of America* 127.4 (2010), pp. 2479–2497.

[22] Robert C Bilger et al. "Standardization of a test of speech perception in noise." In: *Journal of Speech, Language, and Hearing Research* 27.1 (1984), pp. 32–48.

[23] Jens Blauert. *Spatial hearing, revisited edition, the psychophysics of human sound localization.* 1996.

[24] Jens Blauert. *Spatial hearing: the psychophysics of human sound localization.* MIT press, 1997.

[25] S Blausen. "com staff" Medical gallery of Blausen Medical 2014." In: *WikiJournal of Medicine* 1.2 (2014), p. 10.

[26] Robert S Bolia et al. "A speech corpus for multitalker communications research." In: *The Journal of the Acoustical Society of America* 107.2 (2000), pp. 1065–1066.

[27] Ramona Bomhardt. *Anthropometric Individualization of Head-Related Transfer Functions Analysis and Modeling.* Vol. 28. Logos Verlag Berlin GmbH, 2017.

[28] Ramona Bomhardt, Matias de la Fuente Klein, and Janina Fels. "A high-resolution head-related transfer function and three-dimensional ear model database." In: *Proceedings of Meetings on Acoustics 172ASA.* Vol. 29. 1. ASA. 2016, p. 050002.

[29] John S Bradley. "Predictors of speech intelligibility in rooms." In: *The Journal of the Acoustical Society of America* 80.3 (1986), pp. 837–845.

[30] John S Bradley. "Speech intelligibility studies in classrooms." In: *The journal of the acoustical society of America* 80.3 (1986), pp. 846–854.

[31] JS Bradley, RD Reich, and SG Norcross. "On the combined effects of signal-to-noise ratio and room acoustics on speech intelligibility." In: *The Journal of the Acoustical Society of America* 106.4 (1999), pp. 1820–1828.

[32] Thomas Brand et al. "How Do Humans Benefit from Binaural Listening when Recognizing Speech in Noisy and Reverberant Conditions?" In: *Audio Engineering Society Conference: 60th International Conference: DREAMS (Dereverberation and Reverberation of Audio, Music, and Speech).* Audio Engineering Society. 2016.

[33] W Owen Brimijoin, David McShefferty, and Michael A Akeroyd. "Undirected head movements of listeners with asymmetrical hearing impairment during a speech-in-noise task." In: *Hearing research* 283.1-2 (2012), pp. 162–168.

[34] Adelbert W Bronkhorst. "The cocktail party phenomenon: A review of research on speech intelligibility in multiple-talker conditions." In: *Acta Acustica united with Acustica* 86.1 (2000), pp. 117–128.

[35] Adelbert W Bronkhorst. "The cocktail-party problem revisited: early processing and selection of multi-talker speech." In: *Attention, Perception, & Psychophysics* 77.5 (2015), pp. 1465–1487.

[36] Adelbert W Bronkhorst and Reinier Plomp. "A clinical test for the assessment of binaural speech perception in noise." In: *Audiology* 29.5 (1990), pp. 275–285.

[37] AW Bronkhorst and R Plomp. "The effect of head-induced interaural time and level differences on speech intelligibility in noise." In: *The Journal of the Acoustical Society of America* 83.4 (1988), pp. 1508–1516.

[38] AW Bronkhorst and R Plomp. "Binaural speech intelligibility in noise for hearing-impaired listeners." In: *The Journal of the Acoustical Society of America* 86.4 (1989), pp. 1374–1383.

[39] AW Bronkhorst and R Plomp. "Effect of multiple speechlike maskers on binaural speech recognition in normal and impaired hearing." In: *The Journal of the Acoustical Society of America* 92.6 (1992), pp. 3132–3139.

[40] Douglas S Brungart. "Informational and energetic masking effects in the perception of two simultaneous talkers." In: *The Journal of the Acoustical Society of America* 109.3 (2001), pp. 1101–1109.

[41] Douglas S Brungart and Brian D Simpson. "The effects of spatial separation in distance on the informational and energetic masking of a nearby speech signal." In: *The Journal of the Acoustical Society of America* 112.2 (2002), pp. 664–676.

[42] Douglas S Brungart and Brian D Simpson. "Cocktail party listening in a dynamic multitalker environment." In: *Perception & psychophysics* 69.1 (2007), pp. 79–91.

[43] Densil Cabrera, Jianyang Xun, and Martin Guski. "Calculating reverberation time from impulse responses: a comparison of software implementations." In: *Acoustics Australia* 44.2 (2016), pp. 369–378.

[44] Sharon Cameron and Harvey Dillon. "Development of the listening in spatialized noise-sentences test (LISN-S)." In: *Ear and hearing* 28.2 (2007), pp. 196–211.

[45] Sharon Cameron, Harvey Dillon, and Philip Newall. "Development and evaluation of the listening in spatialized noise test." In: *Ear and hearing* 27.1 (2006), pp. 30–42.

[46] Raymond Carhart. "Monaural and binaural discrimination against competing sentences." In: *The Journal of the Acoustical Society of America* 37.6 (1965), pp. 1205–1205.

[47] Raymond Carhart and Tom W Tillman. "Interaction of competing speech signals with hearing losses." In: *Archives of Otolaryngology* 91.3 (1970), pp. 273–279.

[48] Simon Carlile and Virginia Best. "Discrimination of sound source velocity in human listeners." In: *The Journal of the Acoustical Society of America* 111.2 (2002), pp. 1026–1035.

[49] Simon Carlile and Johahn Leung. "The perception of auditory motion." In: *Trends in hearing* 20 (2016), p. 2331216516644254.

[50] Simon Carlile and Daniel Wardman. "Masking produced by broadband noise presented in virtual auditory space." In: *The Journal of the Acoustical Society of America* 100.6 (1996), pp. 3761–3768.

[51] Joseph CK Chan and C Daniel Geisler. "Estimation of eardrum acoustic pressure and of ear canal length from remote points in the canal." In: *The Journal of the Acoustical Society of America* 87.3 (1990), pp. 1237–1247.

[52] Robert B Chaney Jr and JC Webster. "Information in certain multidimensional sounds." In: *The Journal of the Acoustical Society of America* 40.2 (1966), pp. 447–455.

[53] E Colin Cherry. "Some experiments on the recognition of speech, with one and with two ears." In: *The Journal of the acoustical society of America* 25.5 (1953), pp. 975–979.

[54] PRICE CODE. "Sound system equipment–Part 16: Objective rating of speech intelligibility by speech transmission index." In: (2003).

[55] Cynthia L Compton-Conley et al. "Performance of directional microphones for hearing aids: real-world versus simulation." In: *Journal of the American Academy of Audiology* 15.6 (2004), pp. 440–455.

[56] MT Cord, BE Walden, and RM Atack. "Speech recognition in noise test (SPRINT) for H-3 profile." In: *Walter Reed Army Medical Center* (1992).

[57] Stefano Cosentino et al. "A model that predicts the binaural advantage to speech intelligibility from the mixed target and interferer signals." In: *The Journal of the Acoustical Society of America* 135.2 (2014), pp. 796–807.

[58] Robyn M Cox, Genevieve C Alexander, and Christine Gilmore. "Development of the connected speech test (CST)." In: *Ear and Hearing* 8.5 Suppl (1987), 119S–126S.

[59] Robyn M Cox, Ginger A Gray, and Genevieve C Alexander. "Evaluation of a Revised Speech in Noise (RSIN) test." In: *Journal of the American Academy of Audiology* 12.8 (2001), pp. 423–433.

[60] Robyn M Cox et al. "The Connected Speech Test Version 3: Audiovisual Administration." In: *Ear and Hearing* 10.1 (1989), pp. 29–32.

[61] Lothar Cremer and Helmut A Muller. *Principles and Applications of Room Acoustics, Volume 1*. Applied Science Publishers Ltd, 1982.

[62] Herbert Crowley and Roger S Kaufman. "The Rinne tuning fork test." In: *Archives of otolaryngology* 84.4 (1966), pp. 406–408.

[63] JF Culling, M Lavandier, and S Jelfs. "Predicting binaural speech intelligibility in architectural acoustics." In: *The technology of binaural listening*. Springer, 2013, pp. 427–447.

[64] John F Culling. "Speech intelligibility in virtual restaurants." In: *The Journal of the Acoustical Society of America* 140.4 (2016), pp. 2418–2426.

[65] John F Culling and Michael A Akeroyd. "Spatial hearing." In: *The Oxford handbook of auditory science: hearing. Oxford University Press, Oxford* (2010), pp. 123–144.

[66] John F Culling and H Steven Colburn. "Binaural sluggishness in the perception of tone sequences and speech in noise." In: *The Journal of the Acoustical Society of America* 107.1 (2000), pp. 517–527.

[67] John F Culling, Monica L Hawley, and Ruth Y Litovsky. "The role of head-induced interaural time and level differences in the speech reception threshold for multiple interfering sound sources." In: *The Journal of the Acoustical Society of America* 116.2 (2004), pp. 1057–1065.

[68] John F Culling, Kathryn I Hodder, and Chaz Yee Toh. "Effects of reverberation on perceptual segregation of competing voices." In: *The Journal of the Acoustical Society of America* 114.5 (2003), pp. 2871–2876.

[69] John F Culling and Quentin Summerfield. "Measurements of the binaural temporal window using a detection task." In: *The Journal of the Acoustical Society of America* 103.6 (1998), pp. 3540–3553.

[70] BI Dalenbäck and M Strömberg. "Real time walkthrough auralization-the first year." In: *Proceedings of the Institute of Acoustics* 28.2 (2006).

[71] Timothy J Davis, D Wesley Grantham, and René H Gifford. "Effect of motion on speech recognition." In: *Hearing research* 337 (2016), pp. 80–88.

[72] P Dietrich et al. "ITA-Toolbox–An open source Matlab toolbox for acousticians." In: *Fortschritte der Akustik–DAGA* (2012), pp. 151–152.

[73] Donald D Dirks, Donald E Morgan, and Judy R Dubno. "A procedure for quantifying the effects of noise on speech recognition." In: *Journal of Speech and Hearing Disorders* 47.2 (1982), pp. 114–123.

[74] Donald D Dirks and Richard H Wilson. "The effect of spatially separated sound sources on speech intelligibility." In: *Journal of Speech, Language, and Hearing Research* 12.1 (1969), pp. 5–38.

[75] Judy R Dubno, Jayne B Ahlstrom, and Amy R Horwitz. "Binaural advantage for younger and older adults with normal hearing." In: *Journal of Speech, Language, and Hearing Research* 51.2 (2008), pp. 539–556.

[76] Judy R Dubno, Donald D Dirks, and Donald E Morgan. "Effects of age and mild hearing loss on speech recognition in noise." In: *The Journal of the Acoustical Society of America* 76.1 (1984), pp. 87–96.

[77] Judy R Dubno, Amy R Horwitz, and Jayne B Ahlstrom. "Benefit of modulated maskers for speech recognition by younger and older adults with normal hearing." In: *The Journal of the Acoustical Society of America* 111.6 (2002), pp. 2897–2907.

[78] Mark A Ericson. "Multichannel Sound Reproduction in the Environment for Auditory Research." In: *Audio Engineering Society Convention 131*. Audio Engineering Society. 2011.

[79] Gustav Theodor Fechner. *Elemente der Psychophysik: Zweiter Theil*. Breitkopf und Härtel, 1860.

[80] Harald Feldmann. *A History of Audiology: A Comprehensive Report and Bibliography from the Earliest Beginnings to the Present: with 40 Illustrations in the Text*. Beltone Institute for Hearing Research, 1970.

[81] Janina Fels. *Lecture Notes on "Medical Acoustics: Technologies for Hearing Systems and Ultrasound" and "Medical Acoustics: Audiology and Voice"*. Teaching and Research Area of Medical Acoustic Institute of Technical Acoustics RWTH Aachen, 2018.

[82] Selda Fikret-Pasa. "The effect of compression ratio on speech intelligibility and quality." In: *The Journal of the Acoustical Society of America* 95.5 (1994), pp. 2992–2992.

[83] Peter Flipsen Jr. "Measuring the intelligibility of conversational speech in children." In: *Clinical linguistics & phonetics* 20.4 (2006), pp. 303–312.

[84] Norman R French and John C Steinberg. "Factors governing the intelligibility of speech sounds." In: *The journal of the Acoustical society of America* 19.1 (1947), pp. 90–119.

[85] Richard L Freyman et al. "The role of perceived spatial separation in the unmasking of speech." In: *The Journal of the Acoustical Society of America* 106.6 (1999), pp. 3578–3588.

[86] Stanley A Gelfand, Leslie Ross, and Sarah Miller. "Sentence reception in noise from one versus two sources: Effects of aging and hearing loss." In: *The Journal of the Acoustical Society of America* 83.1 (1988), pp. 248–256.

[87] Michael A Gerzon. "Ambisonics in multichannel broadcasting and video." In: *Journal of the Audio Engineering Society* 33.11 (1985), pp. 859–871.

[88] Alessandra Giannela Samelli and Eliane Schochat. "The gaps-in-noise test: gap detection thresholds in normal-hearing young adults." In: *International Journal of Audiology* 47.5 (2008), pp. 238–245.

[89] Helen Glyde et al. "Problems hearing in noise in older adults: a review of spatial processing disorder." In: *Trends in amplification* 15.3 (2011), pp. 116–126.

[90] Helen Glyde et al. "The importance of interaural time differences and level differences in spatial release from masking." In: *The Journal of the Acoustical Society of America* 134.2 (2013), EL147–EL152.

[91] E Bruce Goldstein and James Brockmole. *Sensation and perception*. Cengage Learning, 2016.

[92] Sandra Gordon-Salant. "Hearing loss and aging: new research findings and clinical implications." In: *Journal of Rehabilitation Research & Development* 42 (2005).

[93] Jacques A Grange and John F Culling. "The benefit of head orientation to speech intelligibility in noise." In: *The Journal of the Acoustical Society of America* 139.2 (2016), pp. 703–712.

[94] D Wesley Grantham. "Detection and discrimination of simulated motion of auditory targets in the horizontal plane." In: *The Journal of the Acoustical Society of America* 79.6 (1986), pp. 1939–1949.

[95] D Wesley Grantham and Frederic L Wightman. "Detectability of varying interaural temporal differencesa." In: *The Journal of the Acoustical Society of America* 63.2 (1978), pp. 511–523.

[96] D Wesley Grantham and Frederic L Wightman. "Detectability of a pulsed tone in the presence of a masker with time-varying interaural correlation." In: *The Journal of the Acoustical Society of America* 65.6 (1979), pp. 1509–1517.

[97]   JE Greenberg, PM Peterson, and PM Zurek. "Intelligibility-weighted measures of speech-to-interference ratio and speech system performance." In: *The Journal of the Acoustical Society of America* 94.5 (1993), pp. 3009–3010.

[98]   JJ Groen. "Social hearing handicap: Its measurement by speech audiometry in noise." In: *Int Audiol* 8.1 (1969), pp. 82–183.

[99]   Karl-Heinz Hahlbrock. "Über Sprachaudiometrie und neue Wörterteste." In: *Archiv für Ohren-, Nasen-und Kehlkopfheilkunde* 162.5 (1953), pp. 394–431.

[100] JS Hall. "The development of a new English sentence in noise test and an English number recognition test." In: *Southampton: University of Southampton* (2006).

[101] K Hammer and W Snow. "Binaural Transmission System at Academy of Music in Philadelphia." In: *Memorandum MM-3950, Bell Laboratories* (1932).

[102] J Donald Harris. "Monaural and binaural speech intelligibility and the stereophonic effect based upon temporal cues." In: *The Laryngoscope* 75.3 (1965), pp. 428–446.

[103] William M Hartmann, Brad Rakerd, and Aaron Koller. "Binaural coherence in rooms." In: *Acta acustica united with acustica* 91.3 (2005), pp. 451–462.

[104] Monica L Hawley, Ruth Y Litovsky, and John F Culling. "The benefit of binaural hearing in a cocktail party: Effect of location and type of interferer." In: *The Journal of the Acoustical Society of America* 115.2 (2004), pp. 833–843.

[105] Ira J Hirsh. "The relation between localization and intelligibility." In: *The Journal of the Acoustical Society of America* 22.2 (1950), pp. 196–200.

[106] Merrill Hiscock and Marcel Kinsbourne. "Attention and the right-ear advantage: What is the connection?" In: *Brain and cognition* 76.2 (2011), pp. 263–275.

[107] I Hochmair-Desoyer et al. "The HSM sentence test as a tool for evaluating the speech understanding in noise of cochlear implant users." In: *The American journal of otology* 18.6 Suppl (1997), S83–S83.

[108] Sabine Hochmuth et al. "Influence of noise type on speech reception thresholds across four languages measured with matrix sentence tests." In: *International journal of audiology* 54.sup2 (2015), pp. 62–70.

[109] T Houtgast and H JMi Steeneken. "The modulation transfer function in room acoustics as a predictor of speech intelligibility." In: *Acta Acustica United with Acustica* 28.1 (1973), pp. 66–73.

[110] Tammo Houtgast and Joost M Festen. "On the auditory and cognitive functions that may explain an individual's elevation of the speech reception threshold in noise." In: *International Journal of Audiology* 47.6 (2008), pp. 287–295.

[111] Tammo Houtgast and Herman JM Steeneken. "A review of the MTF concept in room acoustics and its use for estimating speech intelligibility in auditoria." In: *The Journal of the Acoustical Society of America* 77.3 (1985), pp. 1069–1077.

[112] TAMMO Houtgast, Herman JM Steeneken, and R Plomp. "Predicting speech intelligibility in rooms from the modulation transfer function. I. General room acoustics." In: *Acta Acustica united with Acustica* 46.1 (1980), pp. 60–72.

[113] Larry E Humes and Judy R Dubno. "Factors affecting speech understanding in older adults." In: *The aging auditory system*. Springer, 2010, pp. 211–257.

[114] Institute of Technical Acoustics, RWTH Aachen University. *Virtual Acoustics - A real-time auralization framework for scientific research.* http://www.virtualacoustics.org/. Accessed on 2018-03-16. 2018.

[115] American National Standards Institute. *American National Standard: Methods for calculation of the speech intelligibility index.* Acoustical Society of America, 1997.

[116] Master Catalogue-Electronic Instruments. "Brüel & Kjaer." In: *Larsen & Son, Glodstrup, Denmark* (1989).

[117] EN ISO. "3382-1, 2009,"Acoustics—Measurement of Room Acoustic Parameters—Part 1: Performance Spaces,"" in: *International Organization for Standardization, Brussels, Belgium* (2009).

[118] Yukio Iwaya, Yôiti Suzuki, and Daisuke Kimura. "Effects of head movement on front-back error in sound localization." In: *Acoustical science and technology* 24.5 (2003), pp. 322–324.

[119] Sam Jelfs, John F Culling, and Mathieu Lavandier. "Revision and validation of a binaural model for speech intelligibility in noise." In: *Hearing research* 275.1-2 (2011), pp. 96–104.

[120] Morten L Jepsen, Stephan D Ewert, and Torsten Dau. "A computational model of human auditory signal processing and perception." In: *The Journal of the Acoustical Society of America* 124.1 (2008), pp. 422–438.

[121] Gary L Jones and Ruth Y Litovsky. "A cocktail party model of spatial release from masking by both noise and speech interferers." In: *The Journal of the Acoustical Society of America* 130.3 (2011), pp. 1463–1474.

[122] Elke Kalbe et al. "DemTect: a new, sensitive cognitive screening test to support the diagnosis of mild cognitive impairment and early dementia." In: *International journal of geriatric psychiatry* 19.2 (2004), pp. 136–143.

[123] Gavin Kearney et al. "Auditory distance perception with static and dynamic binaural rendering." In: *Audio Engineering Society Conference: 57th International Conference: The Future of Audio Entertainment Technology–Cinema, Television and the Internet.* Audio Engineering Society. 2015.

[124] Gerald Kidd et al. "The role of reverberation in release from masking due to spatial separation of sources for speech identification." In: *Acta acustica united with acustica* 91.3 (2005), pp. 526–536.

[125] Jürgen Kießling. "Moderne Verfahren der Sprachaudiometrie." In: *Laryngo-Rhino-Otologie* 79.11 (2000), pp. 633–635.

[126] Mead C Killion and Patricia A Niquette. "What can the pure-tone audiogram tell us about a patient's SNR loss." In: *Hear J* 53.3 (2000), pp. 46–53.

[127] MEAD C Killion and EDGAR Villchur. "Kessler was right-partly: But SIN test shows some aids improve hearing in noise." In: *Hearing Journal* 46 (1993), pp. 31–31.

[128] Mead C Killion et al. "Development of a quick speech-in-noise test for measuring signal-to-noise ratio loss in normal-hearing and hearing-impaired listeners." In: *The Journal of the Acoustical Society of America* 116.4 (2004), pp. 2395–2405.

[129] Doreen Kimura. "Cerebral dominance and the perception of verbal stimuli." In: *Canadian Journal of Psychology/Revue canadienne de psychologie* 15.3 (1961), p. 166.

[130] Doreen Kimura. "Some effects of temporal-lobe damage on auditory perception." In: *Canadian Journal of Psychology/Revue canadienne de psychologie* 15.3 (1961), p. 156.

[131] Mendel Kleiner, Bengt-Inge Dalenbäck, and Peter Svensson. "Auralization-an overview." In: *Journal of the Audio Engineering Society* 41.11 (1993), pp. 861–875.

[132] S Klockgether and S van de Par. "A model for the prediction of room acoustical perception based on the just noticeable differences of spatial perception." In: *Acta Acustica united with Acustica* 100.5 (2014), pp. 964–971.

[133] Malte Kob and Harald Jers. "Directivity measurement of a singer." In: *The Journal of the Acoustical Society of America* 105.2 (1999), pp. 1003–1003.

[134] WE Kock. "Binaural localization and masking." In: *The Journal of the Acoustical Society of America* 22.6 (1950), pp. 801–804.

[135] Janet Koehnke and Joan M Besing. "A procedure for testing speech intelligibility in a virtual listening environment." In: *Ear and Hearing* 17.3 (1996), pp. 211–217.

[136] W Koenig. "Subjective effects in binaural hearing." In: *The Journal of the Acoustical Society of America* 22.1 (1950), pp. 61–62.

[137] Birger Kollmeier and Robert H Gilkey. "Binaural forward and backward masking: evidence for sluggishness in binaural detection." In: *The Journal of the Acoustical Society of America* 87.4 (1990), pp. 1709–1719.

[138] Birger Kollmeier and Matthias Wesselkamp. "Development and evaluation of a German sentence test for objective and subjective speech intelligibility assessment." In: *The Journal of the Acoustical Society of America* 102.4 (1997), pp. 2412–2421.

[139] Martin Kompis. *Audiologie*. Hans Huber, 2013.

[140] Dan F Konkle and William F Rintelmann. "Introduction to speech audiometry." In: *Principles of Speech Audiometry* (1983), pp. 1–10.

[141] Asbjørn Krokstad, Staffan Strom, and Svein Sørsdal. "Calculating the acoustical room response by the use of a ray tracing technique." In: *Journal of Sound and Vibration* 8.1 (1968), pp. 118–125.

[142] Heinrich Kuttruff. "Stationary propagation of sound energy in flat enclosures with partially diffuse surface reflection." In: *Acta Acustica united with Acustica* 86.6 (2000), pp. 1028–1033.

[143] K Heinrich Kuttruff. "Auralization of impulse responses modeled on the basis of ray-tracing results." In: *Journal of the Audio Engineering Society* 41.11 (1993), pp. 876–880.

[144] Robert M Lambert. "Dynamic theory of sound-source localization." In: *The Journal of the Acoustical Society of America* 56.1 (1974), pp. 165–171.

[145] Norman J Lass and Charles McGregor Woodford. *Hearing science fundamentals*. Elsevier Mosby, 2007.

[146] Howard G Latham. "The signal-to-noise ratio for speech intelligibility—An auditorium acoustics design index." In: *Applied Acoustics* 12.4 (1979), pp. 253–320.

[147] Mathieu Lavandier and John F Culling. "Speech segregation in rooms: Effects of reverberation on both target and interferer." In: *The Journal of the Acoustical Society of America* 122.3 (2007), pp. 1713–1723.

[148] Mathieu Lavandier and John F Culling. "Speech segregation in rooms: Monaural, binaural, and interacting effects of reverberation on target and interferer." In: *The Journal of the Acoustical Society of America* 123.4 (2008), pp. 2237–2248.

[149] Mathieu Lavandier and John F Culling. "Prediction of binaural speech intelligibility against noise in rooms." In: *The Journal of the Acoustical Society of America* 127.1 (2010), pp. 387–399.

[150] Mathieu Lavandier et al. "Binaural prediction of speech intelligibility in reverberant rooms with multiple noise sources." In: *The Journal of the Acoustical Society of America* 131.1 (2012), pp. 218–231.

[151] Minoo Lenarz et al. "Cochlear implant performance in geriatric patients." In: *The Laryngoscope* 122.6 (2012), pp. 1361–1365.

[152] H Levitt and LR Rabiner. "Binaural release from masking for speech and gain in intelligibility." In: *The journal of the acoustical society of america* 42.3 (1967), pp. 601–608.

[153] Harry Levitt and Lawrence R Rabiner. "Use of a sequential strategy in intelligibility testing." In: *The Journal of the Acoustical Society of America* 42.3 (1967), pp. 609–612.

[154] HCCH Levitt. "Transformed up-down methods in psychoacoustics." In: *The Journal of the Acoustical society of America* 49.2B (1971), pp. 467–477.

[155] Liang Li et al. "Does the information content of an irrelevant source differentially affect spoken word recognition in younger and older adults?" In: *Journal of Experimental Psychology: Human Perception and Performance* 30.6 (2004), p. 1077.

[156] JCR Licklider. "The influence of interaural phase relations upon the masking of speech by white noise." In: *The Journal of the Acoustical Society of America* 20.2 (1948), pp. 150–159.

[157] Ruth Y Litovsky. "Spatial release from masking." In: *Acoust. Today* 8.2 (2012), pp. 18–25.

[158] JPA Lochner and JF Burger. "The influence of reflections on auditorium acoustics." In: *Journal of Sound and Vibration* 1.4 (1964), pp. 426–454.

[159] Shannon M Locke, Johahn Leung, and Simon Carlile. "Sensitivity to auditory velocity contrast." In: *Scientific reports* 6 (2016), p. 27725.

[160] J Löhler et al. "Results in using the Freiburger monosyllabic speech test in noise without and with hearing aids." In: *European Archives of Oto-Rhino-Laryngology* 272.9 (2015), pp. 2135–2142.

[161] Micha Lundbeck et al. "Sensitivity to angular and radial source movements as a function of acoustic complexity in normal and impaired hearing." In: *Trends in hearing* 21 (2017), p. 2331216517717152.

[162] Linda Luxon et al. *A Textbook of Audiological Medicine: Clinical Aspects of Hearing and Balance.* CRC Press, 2002.

[163] Sidney A Manning and Edward H Rosenstock. *Classical psychophysics and scaling.* McGraw-Hill New York, 1968.

[164] Nicole Marrone, Christine R Mason, and Gerald Kidd Jr. "The effects of hearing loss and age on the benefit of spatial separation between multiple talkers in reverberant rooms." In: *The Journal of the Acoustical Society of America* 124.5 (2008), pp. 3064–3075.

[165] Bruno Sanches Masiero. *Individualized binaural technology: measurement, equalization and perceptual evaluation.* Vol. 13. Logos Verlag Berlin GmbH, 2012.

[166] Bruno Masiero and Janina Fels. "Perceptually robust headphone equalization for binaural reproduction." In: *Audio Engineering Society Convention 130.* Audio Engineering Society. 2011.

[167] Bruno Masiero, Martin Pollow, and Janina Fels. "Design of a fast broadband individual head-related transfer function measurement system." In: *Acustica, Hirzel* 97 (2011), pp. 136–136.

[168] Rachel A McArdle, Richard H Wilson, and Christopher A Burks. "Speech recognition in multitalker babble using digits, words, and sentences." In: *Journal of the American Academy of Audiology* 16.9 (2005), pp. 726–739.

[169] Sara M Misurelli and Ruth Y Litovsky. "Spatial release from masking in children with normal hearing and with bilateral cochlear implants: Effect of interferer asymmetry." In: *The Journal of the Acoustical Society of America* 132.1 (2012), pp. 380–391.

[170] Henrik Møller. "Fundamentals of binaural technology." In: *Applied acoustics* 36.3-4 (1992), pp. 171–218.

[171]   Henrik Møller et al. "Head-related transfer functions of human subjects."
        In: *Journal of the Audio Engineering Society* 43.5 (1995), pp. 300–321.

[172]   Henrik Møller et al. "Transfer characteristics of headphones measured on
        human ears." In: *Journal of the Audio Engineering Society* 43.4 (1995),
        pp. 203–217.

[173]   Brian CJ Moore. *An introduction to the psychology of hearing*. Brill, 2012.

[174]   Donald E Morgan, Candace A Kamm, and Therese M Velde. "Form
        equivalence of the speech perception in noise (SPIN) test." In: *The Journal
        of the Acoustical Society of America* 69.6 (1981), pp. 1791–1798.

[175]   Masayuki Morimoto and Yoichi Ando. "On the simulation of sound lo-
        calization." In: *Journal of the Acoustical Society of Japan (e)* 1.3 (1980),
        pp. 167–174.

[176]   Dana R Murphy, Meredyth Daneman, and Bruce A Schneider. "Why do
        older adults have difficulty following conversations?" In: *Psychology and
        aging* 21.1 (2006), p. 49.

[177]   GM Naylor. "Treatment of early and late reflections in a hybrid computer
        model for room acoustics." In: *124th ASA meeting*. 1992.

[178]   Michael Nilsson, Sigfrid D Soli, and Jean A Sullivan. "Development of the
        Hearing in Noise Test for the measurement of speech reception thresholds
        in quiet and in noise." In: *The Journal of the Acoustical Society of America*
        95.2 (1994), pp. 1085–1099.

[179]   ICRA Noise Signals. "International collegium of reabilitative audiology."
        In: *ver 0.3* (1997).

[180]   Bertil Nordlund and Björn Fritzell. "The influence of azimuth on speech
        signals." In: *Acta oto-laryngologica* 56.2-6 (1963), pp. 632–642.

[181]   Bertil Nordlund and G Liden. "An artificial head." In: *Acta Oto-
        Laryngologica* 56.2-6 (1963), pp. 493–499.

[182]   Josefa Oberem, Bruno Masiero, and Janina Fels. "Experiments on authen-
        ticity and plausibility of binaural reproduction via headphones employing
        different recording methods." In: *Applied Acoustics* 114 (2016), pp. 71–78.

[183]   Josefa Oberem et al. "Intentional switching in auditory selective atten-
        tion: Exploring different binaural reproduction methods in an anechoic
        chamber." In: *Acta Acustica united with Acustica* 100.6 (2014), pp. 1139–
        1148.

[184] Josefa Oberem et al. "Experiments on localization accuracy with non-individual and individual HRTFs comparing static and dynamic reproduction methods." In: *Fortschritte der Akustik - DAGA 2018*. ISBN 978-3-939296-13-3. Deutsche Gesellschaft fÃ r Akustik e.V. (DEGA). Berlin, 2018.

[185] Wayne O Olsen and Raymond T Carhart. *Development of test procedures for evaluation of binaural hearing aids*. 1967.

[186] Christos Oreinos and Jorg M Buchholz. "Objective analysis of higher-order Ambisonics sound-field reproduction for hearing aid applications." In: *Proceedings of Meetings on Acoustics ICA2013*. Vol. 19. 1. ASA. 2013, p. 055023.

[187] Edward Ozimek et al. "Development and evaluation of Polish digit triplet test for auditory screening." In: *Speech Communication* 51.4 (2009), pp. 307–316.

[188] Edward Ozimek et al. "Speech intelligibility for different spatial configurations of target speech and competing noise source in a horizontal and median plane." In: *Speech Communication* 55.10 (2013), pp. 1021–1032.

[189] Alessia Paglialonga, Gabriella Tognola, and Ferdinando Grandori. "A user-operated test of suprathreshold acuity in noise for adult hearing screening: the SUN (Speech Understanding in Noise) test." In: *Computers in biology and medicine* 52 (2014), pp. 66–72.

[190] Steven van de Par and Armin Kohlrausch. "A new approach to comparing binaural masking level differences at low and high frequencies." In: *The Journal of the Acoustical Society of America* 101.3 (1997), pp. 1671–1680.

[191] Steven van de Par and Armin Kohlrausch. "Dependence of binaural masking level differences on center frequency, masker bandwidth, and interaural parameters." In: *The Journal of the Acoustical Society of America* 106.4 (1999), pp. 1940–1947.

[192] M Torben Pastore and William A Yost. "Spatial Release from Masking with a Moving Target." In: *Frontiers in psychology* 8 (2017), p. 2238.

[193] Stephan Paul. "Binaural recording technology: A historical review and possible future developments." In: *Acta Acustica united with Acustica* 95.5 (2009), pp. 767–788.

[194] Florian Pausch et al. "An Extended Binaural Real-Time Auralization System With an Interface to Research Hearing Aids for Experiments on Subjects With Hearing Loss." In: *Trends in hearing* 22 (2018), p. 2331216518800871.

[195] Jonathan E Peelle et al. "Hearing loss in older adults affects neural systems supporting speech comprehension." In: *Journal of Neuroscience* 31.35 (2011), pp. 12638–12643.

[196] Sönke Pelzer, Marc Aretz, and Michael Vorländer. "Quality assessment of room acoustic simulation tools by comparing binaural measurements and simulations in an optimized test scenario." In: *Proc. Forum Acusticum Aalborg.* 2011.

[197] Sönke Pelzer et al. "Interactive real-time simulation and auralization for modifiable rooms." In: *Building Acoustics* 21.1 (2014), pp. 65–73.

[198] David R Perrott and AD Musicant. "Minimum auditory movement angle: Binaural localization of moving sound sources." In: *The Journal of the Acoustical Society of America* 62.6 (1977), pp. 1463–1466.

[199] M Kathleen Pichora-Fuller and Gurjit Singh. "Effects of age on auditory and cognitive processing: implications for hearing aid fitting and audiologic rehabilitation." In: *Trends in amplification* 10.1 (2006), pp. 29–59.

[200] Christopher J Plack. *The sense of hearing.* Psychology Press, 2013.

[201] R Plomp. "Binaural and monaural speech intelligibility of connected discourse in reverberation as a function of azimuth of a single competing sound source (speech or noise)." In: *Acta Acustica united with Acustica* 34.4 (1976), pp. 200–211.

[202] R Plomp and AM Mimpen. "Improving the reliability of testing the speech reception threshold for sentences." In: *Audiology* 18.1 (1979), pp. 43–52.

[203] R Plomp and AM Mimpen. "Effect of the orientation of the speaker's head and the azimuth of a noise source on the speech-reception threshold for sentences." In: *Acta Acustica United with Acustica* 48.5 (1981), pp. 325–328.

[204] Reinier Plomp. "Auditory handicap of hearing impairment and the limited benefit of hearing aids." In: *The Journal of the Acoustical Society of America* 63.2 (1978), pp. 533–549.

[205] Reinier Plomp. "A signal-to-noise ratio model for the speech-reception threshold of the hearing impaired." In: *Journal of Speech, Language, and Hearing Research* 29.2 (1986), pp. 146–154.

[206] Irwin Pollack. "Binaural Communication Systems: Preliminary Examination." In: *The Journal of the Acoustical Society of America* 31.1 (1959), pp. 81–82.

[207] Luis C Populin. "Monkey sound localization: head-restrained versus head-unrestrained orienting." In: *Journal of Neuroscience* 26.38 (2006), pp. 9820–9832.

[208] Jan-Gerrit Richter, Gottfried Behler, and Janina Fels. "Evaluation of a fast HRTF measurement system." In: *Audio Engineering Society Convention 140*. Audio Engineering Society. 2016.

[209] Donald E Robinson and Lloyd A Jeffress. "Effect of varying the interaural noise correlation on the detectability of tonal signals." In: *The Journal of the Acoustical Society of America* 35.12 (1963), pp. 1947–1952.

[210] EH Rothauser. "IEEE recommended practice for speech quality measurements." In: *IEEE Trans. on Audio and Electroacoustics* 17 (1969), pp. 225–246.

[211] M Rubinstein and L Klein. "The Weber Test: Its Significance in Assessing the True Value of Bone Conduction." In: *Acta oto-laryngologica* 48.3 (1957), pp. 266–275.

[212] Nicholas Schiavetti and RD Kent. "Scaling procedures for the measurement of speech intelligibility." In: *Intelligibility in speech disorders* (1992), pp. 11–34.

[213] Othmar Schimmel et al. "Sound segregation based on temporal envelope structure and binaural cues." In: *The Journal of the Acoustical Society of America* 124.2 (2008), pp. 1130–1145.

[214] Alfred Schmitz. "Ein neues digitales Kunstkopfmeßsystem." In: *Acta Acustica united with Acustica* 81.4 (1995), pp. 416–420.

[215] Alfred Schmitz and Heinrich Bietz. "Free-Field Diffuse-Field Transformation of Artificial Heads." In: *Audio Engineering Society Convention 105*. Audio Engineering Society. 1998.

[216] Bruce A Schneider, Meredyth Daneman, and Dana R Murphy. "Speech comprehension difficulties in older adults: Cognitive slowing or age-related changes in hearing?" In: *Psychology and aging* 20.2 (2005), p. 261.

[217] Dirk Schröder. *Physically based real-time auralization of interactive virtual environments*. Vol. 11. Logos Verlag Berlin GmbH, 2011.

[218] Dirk Schröder and Michael Vorländer. "RAVEN: A real-time framework for the auralization of interactive virtual environments." In: *Forum Acusticum*. Aalborg Denmark. 2011, pp. 1541–1546.

[219] Earl D Schubert. "Some preliminary experiments on binaural time delay and intelligibility." In: *The journal of the acoustical society of america* 28.5 (1956), pp. 895–901.

[220] Maria Schuster et al. "Evaluation of speech intelligibility for children
       with cleft lip and palate by means of automatic speech recognition."
       In: *International Journal of Pediatric Otorhinolaryngology* 70.10 (2006),
       pp. 1741–1747.

[221] Barbara G Shinn-Cunningham et al. "Spatial unmasking of nearby speech
       sources in a simulated anechoic environment." In: *The Journal of the
       Acoustical Society of America* 110.2 (2001), pp. 1118–1129.

[222] Cas Smits, Theo S Kapteyn, and Tammo Houtgast. "Development and
       validation of an automatic speech-in-noise screening test by telephone."
       In: *International journal of audiology* 43.1 (2004), pp. 15–28.

[223] Cas Smits, S Theo Goverts, and Joost M Festen. "The digits-in-noise test:
       assessing auditory speech recognition abilities in noise." In: *The Journal
       of the Acoustical Society of America* 133.3 (2013), pp. 1693–1706.

[224] GA Soulodre, N Popplewell, and John S Bradley. "Combined effects of
       early reflections and background noise on speech intelligibility." In: *Journal
       of Sound and Vibration* 135.1 (1989), pp. 123–133.

[225] Nirmal Kumar Srinivasan, Meghan Stansell, and Frederick J Gallun. "The
       role of early and late reflections on spatial release from masking: Effects of
       age and hearing loss." In: *The Journal of the Acoustical Society of America*
       141.3 (2017), EL185–EL191.

[226] BA Stach. *Clinical Audiology An Introduction-Brad A.* 2010.

[227] Georg Stemmer. *Modeling variability in speech recognition.* Logos-Verlag,
       2005.

[228] I Stemplinger et al. "Zur Verständlichkeit von Einsilbern in unter-
       schiedlichen Störgeräuschen." In: (1994).

[229] Silvanus P Thompson. "LI. On the function of the two ears in the perception
       of space." In: *The London, Edinburgh, and Dublin Philosophical Magazine
       and Journal of Science* 13.83 (1882), pp. 406–416.

[230] Willard R Thurlow, John W Mangels, and Philip S Runge. "Head move-
       ments during sound localization." In: *The Journal of the Acoustical society
       of America* 42.2 (1967), pp. 489–493.

[231] Willard R Thurlow and Philip S Runge. "Effect of induced head movements
       on localization of direction of sounds." In: *The Journal of the Acoustical
       Society of America* 42.2 (1967), pp. 480–488.

[232] Guy Tiberghien. "Initiation à la psychophysique." In: (1984).

[233]   Jerry V Tobias and Jack F Curtis. "Binaural masking of noise by noise." In: *The Journal of the Acoustical Society of America* 31.1 (1959), pp. 127–127.

[234]   Daniel J Tollin et al. "Sound-localization performance in the cat: the effect of restraining the head." In: *Journal of neurophysiology* 93.3 (2005), pp. 1223–1234.

[235]   Laurel Trainor et al. "Development of a flexible, realistic hearing in noise test environment (R-HINT-E)." In: *Signal Processing* 84.2 (2004), pp. 299–309.

[236]   K Tschopp and L Ingold. "Die Entwicklung einer deutschen Version des SPIN-Tests (Speech perception in noise)." In: *Moderne Verfahren der Sprachaudiometrie. Heidelberg: Median-Verlag* (1992), pp. 311–329.

[237]   Michael Vorländer. *Auralization: fundamentals of acoustics, modelling, simulation, algorithms and acoustic virtual reality*. Springer Science & Business Media, 2007.

[238]   K Wagener, T Brand, and B Kollmeier. "Entwicklung und evaluation eines satztests fr die deutsche sprache I: Evaluation des oldenburger satztests." In: *Zeitschrift Audiologie/Audiological Acoustics* 38 (1999a), p. 8695.

[239]   K Wagener, T Brand, and B Kollmeier. "Entwicklung und evaluation eines satztests fr die deutsche sprache II: Evaluation des oldenburger satztests." In: *Zeitschrift Audiologie/Audiological Acoustics* 38 (1999b), p. 8695.

[240]   K Wagener, T Brand, and B Kollmeier. "Entwicklung und evaluation eines satztests fr die deutsche sprache III: Evaluation des oldenburger satztests." In: *Zeitschrift Audiologie/Audiological Acoustics* 38 (1999c), p. 8695.

[241]   Kirsten Carola Wagener and Thomas Brand. "Sentence intelligibility in noise for listeners with normal hearing and hearing impairment: Influence of measurement procedure and masking parameters La inteligibilidad de frases en silencio para sujetos con audición normal y con hipoacusia: la influencia del procedimiento de medición y de los parámetros de enmascaramiento." In: *International journal of audiology* 44.3 (2005), pp. 144–156.

[242]   Nina Wardenga et al. "Do you hear the noise? The German matrix sentence test with a fixed noise level in subjects with normal hearing and hearing impairment." In: *International journal of audiology* 54.sup2 (2015), pp. 71–79.

[243]   Anna Warzybok et al. "Effects of spatial and temporal integration of a single early reflection on speech intelligibility." In: *The Journal of the Acoustical Society of America* 133.1 (2013), pp. 269–282.

[244] Frank Wefers. "A free, open-source software package for directional audio data." In: *Proceedings of the 36th German Annual Conference on Acoustics (DAGA 2010)*. 2010.

[245] Tobias Weissgerber, Tobias Rader, and Uwe Baumann. "Impact of a moving noise masker on speech perception in cochlear implant users." In: *PloS one* 10.5 (2015), e0126133.

[246] Elizabeth M Wenzel et al. "Localization using nonindividualized head-related transfer functions." In: *The Journal of the Acoustical Society of America* 94.1 (1993), pp. 111–123.

[247] Adam Westermann and Jörg M Buchholz. "The effect of spatial separation in distance on the intelligibility of speech in rooms." In: *The Journal of the Acoustical Society of America* 137.2 (2015), pp. 757–767.

[248] Adam Westermann and Jörg M Buchholz. "The effect of nearby maskers on speech intelligibility in reverberant, multi-talker environments." In: *The Journal of the Acoustical Society of America* 141.3 (2017), pp. 2214–2223.

[249] Felix A Wichmann and N Jeremy Hill. "The psychometric function: I. Fitting, sampling, and goodness of fit." In: *Perception & psychophysics* 63.8 (2001), pp. 1293–1313.

[250] Felix A Wichmann and N Jeremy Hill. "The psychometric function: II. Bootstrap-based confidence intervals and sampling." In: *Perception & psychophysics* 63.8 (2001), pp. 1314–1329.

[251] Fred Wightman and Doris Kistler. "Measurement and validation of human HRTFs for use in hearing research." In: *Acta acustica united with Acustica* 91.3 (2005), pp. 429–439.

[252] Frederic L Wightman and Doris J Kistler. "Headphone simulation of free-field listening. I: stimulus synthesis." In: *The Journal of the Acoustical Society of America* 85.2 (1989), pp. 858–867.

[253] Frederic L Wightman and Doris J Kistler. "Resolution of front–back ambiguity in spatial hearing by listener and source movement." In: *The Journal of the Acoustical Society of America* 105.5 (1999), pp. 2841–2853.

[254] Richar H Wilson and Rachel McArdle. "Speech signals used to evaluate functional status of the auditory system." In: *Journal of Rehabilitation Research & Development* 42 (2005).

[255] Richard H Wilson. "Development of a speech-in-multitalker-babble paradigm to assess word-recognition performance." In: *Journal of the American Academy of Audiology* 14.9 (2003), pp. 453–470.

[256] Richard H Wilson, Harvey B Abrams, and Amanda L Pillion. "A word-recognition task in multitalker babble using a descending presentation mode from 24 dB to 0 dB signal to babble." In: *Journal of Rehabilitation Research and Development* 40.4 (2003), pp. 321–328.

[257] Richard H Wilson and Rachel McArdle. "Intra-and inter-session test, retest reliability of the Words-in-Noise (WIN) test." In: *Journal of the American Academy of Audiology* 18.10 (2007), pp. 813–825.

[258] Richard H Wilson, Rachel A McArdle, and Sherri L Smith. "An evaluation of the BKB-SIN, HINT, QuickSIN, and WIN materials on listeners with normal hearing and listeners with hearing loss." In: *Journal of Speech, Language, and Hearing Research* 50.4 (2007), pp. 844–856.

[259] Richard H Wilson et al. "Measurements of auditory thresholds for speech stimuli." In: *Principles of speech audiometry* (1983), pp. 79–126.

[260] William A Yost. "Spatial release from masking based on binaural processing for up to six maskers." In: *The Journal of the Acoustical Society of America* 141.3 (2017), pp. 2093–2106.

[261] Pavel Zahorik. "Direct-to-reverberant energy ratio sensitivity." In: *The Journal of the Acoustical Society of America* 112.5 (2002), pp. 2110–2117.

[262] Pavel Zahorik. "Perceptually relevant parameters for virtual listening simulation of small room acoustics." In: *The Journal of the Acoustical Society of America* 126.2 (2009), pp. 776–791.

[263] Pavel Zahorik, Douglas S Brungart, and Adelbert W Bronkhorst. "Auditory distance perception in humans: A summary of past and present research." In: *ACTA Acustica united with Acustica* 91.3 (2005), pp. 409–420.

[264] Peisheng Zhu et al. "Objective Rating Methods of Speech Intelligibility-The Interpretation on the IEC 60268-16 Standard." In: *Diansheng Jishu(Audio Engineering)* 36.5 (2012), pp. 40–45.

[265] Melanie A Zokoll et al. "Internationally comparable screening tests for listening in noise in several European languages: The German digit triplet test as an optimization prototype." In: *International Journal of Audiology* 51.9 (2012), pp. 697–707.

[266] Patrick M Zurek, Richard L Freyman, and Uma Balakrishnan. "Auditory target detection in reverberation." In: *The Journal of the Acoustical Society of America* 115.4 (2004), pp. 1609–1620.

[267] Eberhard Zwicker and Hugo Fastl. *Psychoacoustics: Facts and models.* Vol. 22. Springer Science & Business Media, 2013.

Curriculum Vitae

Personal Data

	Rhoddy Angel Viveros Muñoz
14.08.1981	born in Linares, Chile

Education

| 1996–1999 | Liceo Juan Ignacio Molina, Linares |

Higher Education

| 2003–2009 | Master of Science in Electrical Engineering |
| | Pontificia Universidad Catolica de Valparaiso, Chile |

Work Experience

2014 – 2019	Research Assistant
	Institute of Technical Acoustics, Medical Acoustics Group
	RWTH Aachen University
2010 – 2014	Electrical Project Engineer
	Pares & Alvares Ingeniería y Proyectos
	Concepcion, Chile
2009 – 2010	Electrical Engineer
	Bettoli SA
	Concepcion, Chile

July 18, 2019

Bisher erschienene Bände der Reihe

Aachener Beiträge zur Akustik

ISSN 1866-3052
ISSN 2512-6008 (seit Band 28)

1   Malte Kob   Physical Modeling of the Singing Voice

                          ISBN 978-3-89722-997-6   40.50 EUR

2   Martin Klemenz   Die Geräuschqualität bei der Anfahrt elektrischer
                          Schienenfahrzeuge
                          ISBN 978-3-8325-0852-4   40.50 EUR

3   Rainer Thaden   Auralisation in Building Acoustics

                          ISBN 978-3-8325-0895-1   40.50 EUR

4   Michael Makarski   Tools for the Professional Development of Horn
                          Loudspeakers
                          ISBN 978-3-8325-1280-4   40.50 EUR

5   Janina Fels   From Children to Adults: How Binaural Cues and
                          Ear Canal Impedances Grow
                          ISBN 978-3-8325-1855-4   40.50 EUR

6   Tobias Lentz   Binaural Technology for Virtual Reality

                          ISBN 978-3-8325-1935-3   40.50 EUR

7   Christoph Kling   Investigations into damping in building acoustics
                          by use of downscaled models
                          ISBN 978-3-8325-1985-8   37.00 EUR

8   Joao Henrique Diniz   Modelling the dynamic interactions of rolling
    Guimaraes   bearings
                          ISBN 978-3-8325-2010-6   36.50 EUR

9   Andreas Franck   Finite-Elemente-Methoden, Lösungsalgorithmen und
                          Werkzeuge für die akustische Simulationstechnik
                          ISBN 978-3-8325-2313-8   35.50 EUR

10 Sebastian Fingerhuth   Tonalness and consonance of technical sounds

ISBN 978-3-8325-2536-1   42.00 EUR

11 Dirk Schröder   Physically Based Real-Time Auralization of Interactive Virtual Environments
ISBN 978-3-8325-2458-6   35.00 EUR

12 Marc Aretz   Combined Wave And Ray Based Room Acoustic Simulations Of Small Rooms
ISBN 978-3-8325-3242-0   37.00 EUR

13 Bruno Sanches Masiero   Individualized Binaural Technology. Measurement, Equalization and Subjective Evaluation
ISBN 978-3-8325-3274-1   36.50 EUR

14 Roman Scharrer   Acoustic Field Analysis in Small Microphone Arrays

ISBN 978-3-8325-3453-0   35.00 EUR

15 Matthias Lievens   Structure-borne Sound Sources in Buildings

ISBN 978-3-8325-3464-6   33.00 EUR

16 Pascal Dietrich   Uncertainties in Acoustical Transfer Functions. Modeling, Measurement and Derivation of Parameters for Airborne and Structure-borne Sound
ISBN 978-3-8325-3551-3   37.00 EUR

17 Elena Shabalina   The Propagation of Low Frequency Sound through an Audience
ISBN 978-3-8325-3608-4   37.50 EUR

18 Xun Wang   Model Based Signal Enhancement for Impulse Response Measurement
ISBN 978-3-8325-3630-5   34.50 EUR

19 Stefan Feistel   Modeling the Radiation of Modern Sound Reinforcement Systems in High Resolution
ISBN 978-3-8325-3710-4   37.00 EUR

20 Frank Wefers   Partitioned convolution algorithms for real-time auralization
ISBN 978-3-8325-3943-6   44.50 EUR

21 Renzo Vitale   Perceptual Aspects Of Sound Scattering In
                            Concert Halls
                            ISBN 978-3-8325-3992-4   34.50 EUR

22 Martin Pollow   Directivity Patterns for Room Acoustical
                            Measurements and Simulations
                            ISBN 978-3-8325-4090-6   41.00 EUR

23 Markus Müller-Trapet   Measurement of Surface Reflection
                            Properties. Concepts and Uncertainties
                            ISBN 978-3-8325-4120-0   41.00 EUR

24 Martin Guski   Influences of external error sources on
                            measurements of room acoustic parameters
                            ISBN 978-3-8325-4146-0   46.00 EUR

25 Clemens Nau   Beamforming in modalen Schallfeldern von
                            Fahrzeuginnenräumen
                            ISBN 978-3-8325-4370-9   47.50 EUR

26 Samira Mohamady   Uncertainties of Transfer Path Analysis
                            and Sound Design for Electrical Drives
                            ISBN 978-3-8325-4431-7   45.00 EUR

27 Bernd Philippen   Transfer path analysis based on in-situ
                            measurements for automotive applications
                            ISBN 978-3-8325-4435-5   50.50 EUR

28 Ramona Bomhardt   Anthropometric Individualization of Head-Related
                            Transfer Functions Analysis and Modeling
                            ISBN 978-3-8325-4543-7   35.00 EUR

29 Fanyu Meng   Modeling of Moving Sound Sources
                            Based on Array Measurements
                            ISBN 978-3-8325-4759-2   44.00 EUR

30 Jan-Gerrit Richter   Fast Measurement of Individual Head-Related
                            Transfer Functions
                            ISBN 978-3-8325-4906-0   45.50 EUR

31 Rhoddy A. Viveros   Speech perception in complex acoustic environments:
    Muñoz   Evaluating moving maskers using virtual acoustics
                            ISBN 978-3-8325-4963-3   35.50 EUR

Alle erschienenen Bücher können unter der angegebenen ISBN-Nummer direkt online
(http://www.logos-verlag.de) oder per Fax (030 - 42 85 10 92) beim Logos Verlag
Berlin bestellt werden.